Lecture Notes of the Institute for Computer Sciences, Social Informatics and Telecommunications Engineering 615

The LNICST series publishes ICST's conferences, symposia and workshops.

LNICST reports state-of-the-art results in areas related to the scope of the Institute.

The type of material published includes

- Proceedings (published in time for the respective event)
- Other edited monographs (such as project reports or invited volumes)

LNICST topics span the following areas:

- General Computer Science
- E-Economy
- E-Medicine
- Knowledge Management
- Multimedia
- Operations, Management and Policy
- Social Informatics
- Systems

Xiuhua Li · Victor C. M. Leung · Xiaofei Wang
Editors

6GN for Future Wireless Networks

7th EAI International Conference, 6GN 2024
Chongqing, China, October 19–20, 2024
Proceedings

Springer

Editors
Xiuhua Li
Chongqing University
Chongqing, China

Victor C. M. Leung
The University of British Columbia
Vancouver, BC, Canada

Xiaofei Wang
Tianjin University
Tianjin, China

ISSN 1867-8211 ISSN 1867-822X (electronic)
Lecture Notes of the Institute for Computer Sciences, Social Informatics and Telecommunications Engineering
ISBN 978-3-031-93192-5 ISBN 978-3-031-93193-2 (eBook)
https://doi.org/10.1007/978-3-031-93193-2

This Springer imprint is published by the registered company Springer Nature Switzerland AG
The registered company address is: Gewerbestrasse 11, 6330 Cham, Switzerland

Preface

We are delighted to introduce the proceedings of the seventh edition of the European Alliance for Innovation (EAI) International Conference on 6G for Future Wireless Networks (6GN 2024). This conference brought together researchers, developers and practitioners around the world who are leveraging and developing 6G key technology enablers, 6G use cases and artificial intelligence for 6G. The theme of 6GN 2024 was "6GN Inspired Future Technologies: A Smarter World".

The technical program of 6GN 2024 consisted of 18 full papers in oral presentation sessions at the main conference tracks. The conference tracks were: Main Track, Late Track and Last-Minute Track. The conference sessions were: Session 1, Wireless Control and Network Optimization; Session 2, Edge Computing and Resource Allocation; Session 3, Intelligent Systems and Simulation; Session 4: Artificial Intelligence and Digital Economy.

Coordination with the steering committee, Shuo Shi and Ao Li, was essential for the success of the conference. We sincerely appreciate their constant support and guidance. It was also a great pleasure to work with such an excellent organizing committee team for their hard work in organizing and supporting the conference. In particular, the Technical Program Committee, led by our TPC Co-Chairs, Xin Liu, Mu Zhou, Feng Li and Qilin Fan, completed the peer-review process of technical papers and made a high-quality technical program. We are also grateful to Conference Manager Ivana Bujdakova for her support and to all the authors who submitted their papers to the 6GN 2024 conference.

We strongly believe that the 6GN conference provides a good forum for all researchers, developers and practitioners to discuss all science and technology aspects that are relevant to 6GN. We also expect that future 6GN conferences will be as successful and stimulating as indicated by the contributions presented in this volume.

December 2024

Xiuhua Li
Victor C. M. Leung
Xiaofei Wang

Organization

Steering Committee

Shuo Shi	Harbin Institute of Technology, China
Ao Li	Harbin University of Science and Technology, China

Organizing Committee

General Chair

Xiuhua Li	Chongqing University, China

General Co-chairs

Victor C. M. Leung	University of British Columbia, Canada
Xiaofei Wang	Tianjin University, China

TPC Chair and Co-chairs

Xin Liu	Dalian University of Technology, China
Mu Zhou	Chongqing University of Posts and Telecommunications, China
Feng Li	Zhejiang Gongshang University, China
Qilin Fan	Chongqing University, China

Sponsorship and Exhibit Chair

Wei Zhou	Chongqing University, China

Local Chairs

Qilin Fan	Chongqing University, China
Ke Yu	Chongqing University, China

Publications Chair

Ke Yu	Chongqing University, China

Technical Program Committee

Chunqiang Hu	Chongqing University, China
Qilin Fan	Chongqing University, China
Shu Fu	Chongqing University, China
Chao Qiu	Tianjin University, China
Ran Wang	Nanjing University of Aeronautics and Astronautics, China
Xiaolong Yang	Chongqing University of Posts and Telecommunications, China
Li Zhou	National University of Defense Technology, China
Fang Fang	Western University, Canada
Ke Yu	Chongqing University, China
Ying He	Shenzhen University, China
Huiwei Wang	Southwest University, China
Ru Huo	Beijing University of Technology, China
Hui Li	Chongqing University, China

Contents

Wireless Control and Network Optimization

Analysis of Electronic Countermeasures and Exploration of Cognitive Techniques 3
Jialei Sun, Xi Wang, Chenyu Li, and Shuo Shi

Base Station Location Optimization Framework Based on Digital Twins for 6G Networks 14
Zongxuan Sha, Ru Huo, and Tao Huang

Modeling and Optimization of Random Access in NB-IoT Network with Heterogeneous Users 28
Haiheng Ye and Peiran Wu

Research on Routing Algorithm of Space-Air Integrated Networks 45
Kexin Zhang, Zhongyu Yin, Tong Liu, and Shuo Shi

Research on Statistical Priority-Based Multiple Access Protocol in Mobile Ad Hoc Networks 55
Han Li, Shouming Wei, Cheng Zhang, and Shuo Shi

Edge Computing and Resource Allocation

Deep Reinforcement Learning for Cache Allocation of Multi-cloud Content Providers in Mobile Edge-Cloud Computing Networks 71
Gang Wang, Jinlong Hao, Guozeng Xu, Jian Tang, Xiuhua Li, and Penghua Li

HCTRL: Hierarchical Clustering Tree Reinforcement Learning Algorithm for Efficient Content Caching 83
Ruohan Shi, Zihong Ming, Qilin Fan, Xiuhua Li, Kai Wang, and Xiang Deng

Multivariate Time Series Anomaly Detection Based on Device-Edge-Cloud Collaboration in Internet of Things 98
Yinkang Xu, Haowei Li, Zhiying Xiong, Qilin Fan, Xiuhua Li, and Cheng Zhang

Intelligent Systems and Simulation

UAV Topology Control Method Based on Fuzzy C-Clustering Algorithm 113
Jiajie Guo, Jian He, Dawei Chen, and Shuo Shi

Research on Joint Optimization Model of Drone Trajectory and Power for SAGIN Swarm Intelligence Technology 125
Zhenyu Xu, XiuLian Lin, and XinLu Li

Research on the Application of LoRa Technology in Intelligent Ranch Environmental Monitoring System 134
Yanhong Zhang, Zhenyu Xu, and Jingzhong Li

Simulation and Analysis of Harmonic Reducer Dynamics Based on Adams 143
Fan Zhenduo

V2V Communication Systems for Advanced Truck Platooning at Freight Hubs 153
Tao Song, Jinlong Hao, Luxi Cheng, Hui Li, Xiuhua Li, and Penghua Li

Artificial Intelligence and Digital Economy

A Software-Hardware Co-design Platform for Convolutional Neural Network Accelerator with Tengine 169
Zhiwei Liu, Shangwen Sui, Wanyu Lu, Bin Yu, and Ao Li

Personalized Federated Learning Algorithm Based on Shuffle Differential Privacy 179
Ruifeng Zhao, Jiangang Lu, Zufeng Hou, Yong Zhang, Chao Wang, Jiantian Chen, Kaiyan Pan, and Haixin Liu

Research and Design of Tomato Detection Accelerator Based on FPGA 191
Bin Yu, Wanyu Lu, Shangwen Sui, Zhiwei Liu, and Ao Li

Research on the Role of Fiscal Deficit in the Impact of Digital Inclusive Finance on Economic Growth 199
Xiaofeng Zhang and Fangfang Ji

The Impact of E-Commerce on the Consumption Level of Chinese Residents Under the Background of Communication Network Technology 209
Xiaofeng Zhang and Weiyun Song

Author Index 219

Wireless Control and Network Optimization

Analysis of Electronic Countermeasures and Exploration of Cognitive Techniques

Jialei Sun, Xi Wang, Chenyu Li, and Shuo Shi(✉)

Harbin Institute of Technology, Harbin 150001, Heilongjiang, China
crcss@hit.edu.cn

Abstract. Ad Hoc networks are self-organized by wireless mobile nodes using distributed protocols without relying on central nodes or fixed communication infrastructures. While disrupting adversarial communication systems might offer strategic advantages by undermining their communicative efficacy, traditional interdiction approaches often fall short in the context of unfamiliar Ad Hoc networks. This project aims to conduct an exhaustive exploration of the distinctive properties and technical attributes of Ad Hoc networks. It appraises the impacts posed by conventional electronic countermeasures on their communication flows, delineating both the potential and the intrinsic limitations of these methods. Recognizing the constraints of traditional tactics, the research integrates advanced cognitive electronic warfare strategies with robust anti-interference techniques, leveraging IEEE802.11 fallback mechanisms within Ad Hoc networks. This initiative facilitates the development of innovative store-and-replay methodologies, capable of inflicting targeted disturbances or security breaches at the network tier. These explorations provide critical insights for countering nascent challenges and significantly bolster the fortitude to withstand and counteract interference in unknown communication networks, thereby securing strategic superiority and initiative on the battlefield.

Keywords: Ad Hoc · Electronic Countermeasures · Cognitive Electronic Warfare Techniques

1 Introduction

Ad Hoc network is a type of wireless network made up of wireless mobile nodes that communicate using a distributed protocol, eliminating the need for a central node [1]. This network operates independently of fixed communication infrastructure, offering benefits such as ease of setup, versatility in different environments, resilience against disruptions, support for mobile nodes, and cost-effectiveness [2]. Due to these advantages, military forces have adopted Ad Hoc networks for mobile communications in battlefield settings. These networks form the technological backbone and security framework for portable airborne combat systems being developed and used by various countries.

Effectively disrupting an enemy's communication systems can significantly undermine their communication capabilities, granting us a vital strategic advantage [3]. This

X. Li et al. (Eds.): 6GN 2024, LNICST 615, pp. 3–13, 2025.
https://doi.org/10.1007/978-3-031-93193-2_1

advantage is crucial for securing and maintaining battlefield initiative. Currently, various network attack methods exist, such as routing black hole attacks and wormhole attacks [4]. These techniques are typically used to exploit vulnerabilities in the standard Internet link layer or network layer protocols. However, they heavily rely on prior knowledge of the network, making traditional computer network warfare techniques ineffective when there is insufficient information about the enemy's communication systems. Consequently, computer network warfare often becomes synonymous with communication electronic warfare, where electronic jamming serves as the primary method of disruption [5]. Thus, it is essential to assess the specific effectiveness of electronic jamming techniques in Ad Hoc networks. Traditional jamming methods have their pros and cons, highlighting the urgent need to explore innovative attack strategies against unknown Ad Hoc networks. This exploration is crucial for maintaining our information superiority and effectively countering potential threats from adversaries.

2 Ad Hoc Characteristics and Technical Features

Ad Hoc network is a novel wireless network distinguished by its decentralized, self-organizing, and multi-hop transmission features. Comprising a collection of wireless terminal devices, each device functions autonomously as a network node, equipped with wireless transceivers and routing capabilities. These nodes possess the ability to move freely, establish communication links spontaneously, and dynamically adjust their configurations as needed [6].

Distributed characteristics: In Ad Hoc networks, all communication nodes share equal status, relying on a unified distributed protocol for interoperability. Consequently, there is no central node in the traditional sense.

Dynamic topology characteristics: The topology of Ad Hoc networks is highly dynamic due to the ability of user terminal equipment to power on or off and move freely. These movements result in changes in the number and location of nodes within the network, leading to continuous changes in the network's overall topology.

Multi-hop communication characteristics: Terminal devices in Ad Hoc networks are often mobile and constrained by portable energy, limiting the transmit power of individual nodes. Consequently, the communication coverage of a single node is restricted, necessitating multi-hop routing and forwarding through intermediate nodes to facilitate information exchange between distant parties within the network.

Channel quality: Unlike wired networks that rely on infrastructure support, Ad Hoc networks depend entirely on the wireless transceiver capabilities of individual terminal devices for communication [7].

3 Ad Hoc Electronic Countermeasures Analysis

Ad Hoc networks employ various implementation methods depending on their intended applications, each utilizing specific protocols, message formats, and other elements. However, the absence of standardized protocols results in a lack of uniformity across different networks. Additionally, the physical layer of these networks may incorporate techniques such as frequency hopping, spread spectrum, and time hopping. Notably, the primary method of network countermeasures is communications electronic warfare rather than computer network warfare. Despite advancements, radio communication remains pivotal for interconnection and collaborative command within these networks [8]. Consequently, wherever there are exposed radio signals, conditions conducive to implementing communication jamming exist. Thus, an analysis of typical communication jamming methods is warranted.

In the analysis of communication jamming, we introduce the jamming coefficient to evaluate the impact of jamming signals under varying conditions. For the communication nodes within an Ad Hoc network, this coefficient is defined as the ratio of the jamming signal power P_j at the input of the receiving node to the power of the interfered signal P_i when effective jamming occurs. It is influenced by factors such as transmit power, antenna gain, distance, and correction factor, collectively determining the extent of jamming on the target communication. The jamming coefficient is denoted as:

$$\rho_i = \frac{P_j \cdot G_{jr} \cdot G_{rj} \cdot R_{rs}^2}{P_i \cdot G_{ir} \cdot G_{ri} \cdot R_{rj}^2} \cdot \frac{p}{L_{FH} \cdot L_{Window} \cdot L_{ccsk}} \tag{1}$$

Here, G_{jr} and G_{rj} represent the antenna gains from the jamming signal source to the receiver and from the receiver to the jamming signal source, respectively. Similarly, G_{ir} and G_{ri} denote the antenna gains from the jammed source to the receiver and from the receiver to the jammed source, respectively. R_{rs} and R_{rj} denote the distances from the receiver to the target source and from the receiver to the jamming source, respectively. The correction factors encompass losses such as frequency hopping loss L_{FH}, window loss L_{Window}, and other specific losses L_{ccsk}.

3.1 Targeted Jamming

Targeted jamming occurs when the carrier frequency of the jamming aligns with the signal frequency. As depicted in Fig. 1, the spectrum of targeted jamming shows a concentrated power distribution with a narrower jamming band. In this method, all jamming energy is directed towards suppressing the enemy's communication signals, resulting in a high jamming power utilization rate. However, achieving effective targeted jamming demands precise frequency alignment, necessitating advanced performance capabilities from the jamming equipment. Additionally, reconnaissance is essential to guide the selection of jamming frequencies.

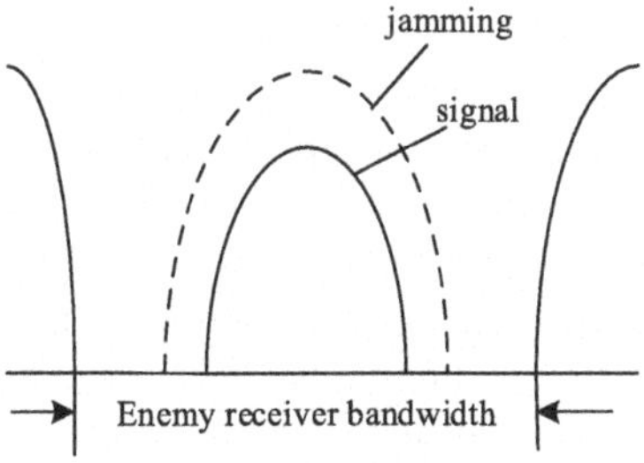

Fig. 1. Schematic diagram of targeting jamming.

3.2 Semi- targeted Jamming

Semi-targeted jamming is characterized by less precision or incomplete frequency alignment compared to targeted jamming. Typically, the spectrum of the jamming signal is broader than the bandwidth of the suppressed enemy communication signal. While the jamming spectrum may encompass all or most of the frequencies selected by the enemy receiver, even if there is not a high degree of frequency coincidence with the enemy signal spectrum, it can still generate a certain level of jamming. The drawback of semi-targeted jamming lies in its lack of power concentration, resulting in a lower utilization rate, applicable only in specific circumstances. For instance, when enemy signals emerge within a brief timeframe, making precise targeting impractical, or when certain communication modes do not necessitate precise frequency alignment, semi-targeted jamming can still yield effective results.

3.3 Continuous Spectrum Blocking Jamming

Continuous spectrum blocking jamming is a prevalent mode of suppressive jamming, particularly observed in Ad Hoc networks with broad bandwidth or employing frequency hopping technology. This jamming mode specifically targets frequency hopping across all bands following the initial signal blockage [9]. Consequently, communication signals experience jamming regardless of their frequency hopping patterns, effectively suppressing their transmission.

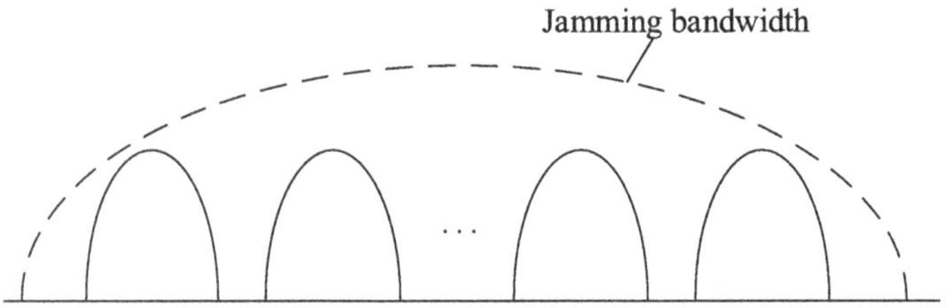

Fig. 2. Schematic diagram of continuous spectrum blocking jamming.

Figure 2, depicting the schematic of continuous spectrum blocking jamming, it's evident that this jamming type is uniformly distributed across the signal band. As the number of signal frequency points increases, the corresponding coverage bandwidth should also expand.

Figure 3, depicting the schematic of broadband blocking jamming scenarios and technical requirements. For full-channel blocking jamming in an Ad Hoc network with an operational bandwidth of 237 MHz, effective jamming can be achieved by ensuring that the noise jamming signal covers this entire band. However, if the network employs spread spectrum and RS code error correction, most of the jamming signal energy becomes ineffective, necessitating a very high signal-to-noise ratio for effective jamming. Based on calculations, a bit error rate (BER) occurs when the signal-to-noise ratio reaches 23 dB, while a BER of 0.01 requires a ratio of 28 dB, and a BER of 0.1 necessitates a ratio of 33 dB. If the jammer were to employ an omni-directional antenna for jamming transmissions, positioned at the same distance of 200 km ahead of the enemy fleet, the average transmit power required to induce BERs in the Ad Hoc network would need to reach approximately 210 kilowatts.

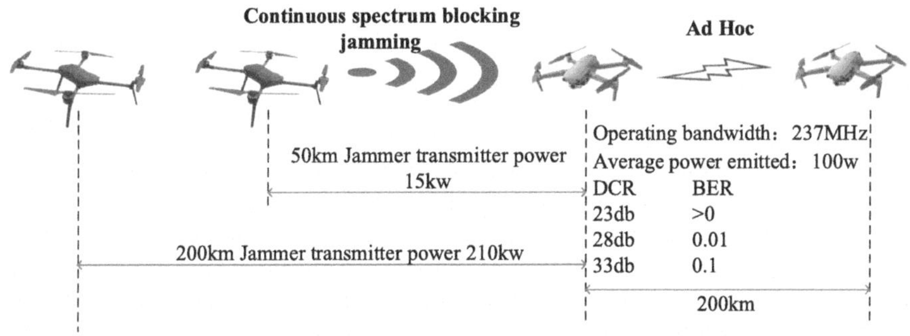

Fig. 3. Schematic diagram of broadband blocking jamming scenarios and technical requirements.

This type of jamming offers the advantage of not requiring specific knowledge about the signal's frequency hopping rate or frequency information. Simply covering the signal band entirely can achieve jamming, making it relatively straightforward. However, ensuring an effective jamming effect necessitates sufficient jamming power, and the distribution of signals across the band may lead to some bandwidth being wasted. To minimize this wastage, particularly when significant jamming power is needed, strategies such as positioning jammers closer to the enemy receiver, increasing the number of jammers, or reducing jamming bandwidth can be employed.

3.4 Partial-Band Interdiction Jamming

Partial-band blocking jamming is blocking communication signals on some sub bands of Ad Hoc network, and its jamming schematic is shown in Fig. 4. When the communication band is changed, the corresponding coverage bandwidth should be changed as well.

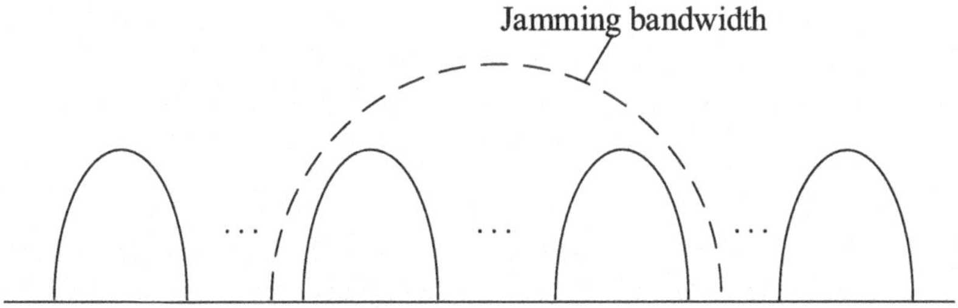

Fig. 4. Schematic diagram of partial-band blocking jamming.

In comparison to continuous spectrum blocking jamming, semi-targeted jamming exhibits a narrower jamming frequency range, resulting in more concentrated jamming power and a better jamming effect within the targeted bands. However, its limitations should not be overlooked. Semi-targeted jamming can only jam with specific frequency bands, leaving signals in other bands unaffected. Therefore, it necessitates that the jamming party possesses a certain level of prior knowledge about the signals to be jammed with in order to achieve effective jamming.

3.5 Comb-Spectrum Blocking Jamming

As its name suggests, comb spectrum blocking jamming presents itself as a comb-like spectrum within the jamming signal's bandwidth. Specifically, it entails setting specific frequencies as center frequencies and establishing intervals between frequencies, each with its designated bandwidth for blocking [10]. This approach represents a form of spectrum-discontinuous targeted blocking jamming. It enhances jamming power concentration, enabling more efficient utilization of jamming power, and reduces the dispersion of jamming bandwidth into independent frequency points. However, its high jamming efficiency comes with increased operational complexity. Implementing this type of jamming requires obtaining information about the communication band being jammed with and details about the frequency hopping intervals, as depicted in Fig. 5.

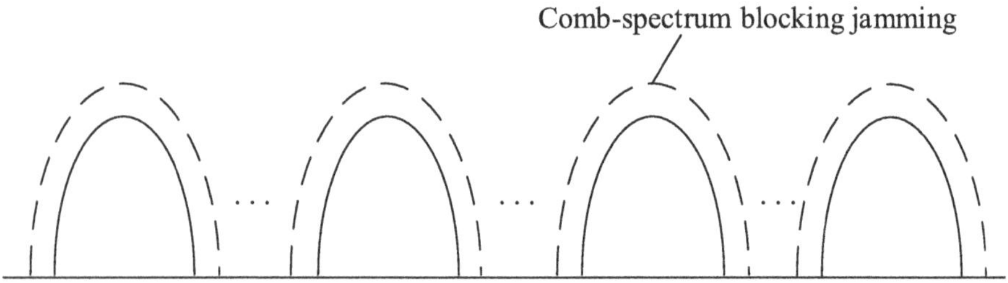

Fig. 5. Schematic diagram of trace spectrum blocking jamming

In an Ad Hoc network operating with 51 frequency points, generating comb jamming that covers all these points results in a signal-to-noise ratio of 10 dB, leading to a BER. Increasing the signal-to-noise ratio to 17 dB reduces the BER to 0.1. Consequently, the power requirement for the jammer positioned 200 km away decreases to 10.4 kW, and when the BER reaches 0.1, the jamming power decreases to 52.4 kW. Similarly, at 50 km, the required jammer power is 0.7 kW and 3.3 kW respectively.

3.6 Repeater Jamming

Repeater jamming adjusts its jamming frequency according to the working frequency of the target communication system [11], primarily focusing on frequency-hopping systems and other frequency-changing communication systems. When the jamming frequency adjustment speed matches that of the target communication system, this type of jamming can precisely disrupt the communication link, leading to efficient utilization of jamming power. However, the practical implementation of such jamming is challenging due to the constraints imposed by the jamming ellipse relationship.

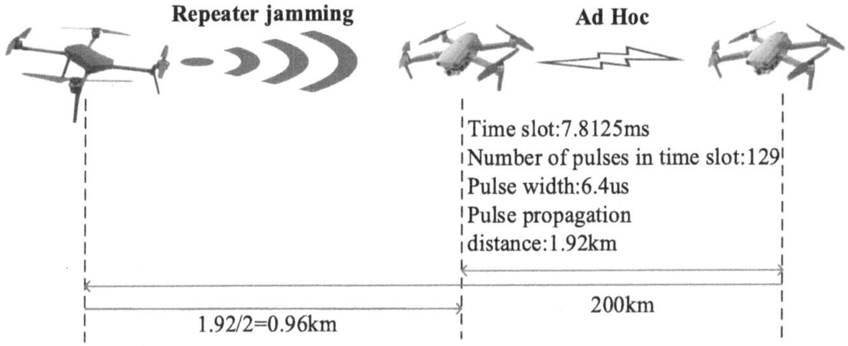

Fig. 6. Schematic repeater jamming scenarios and technical requirements.

Figure 6, depicting the schematic of forward jamming scenarios and technical requirements. For the Ad Hoc network using TDMA for time slot allocation, if the basic allocation time slot is 7.8125 ms, in each time slot, the Ad Hoc network using the double-pulse communication mode can transmit 129 pulses, and the width of each pulse is 6.4us. Such a narrow pulse width with the inter-pulse hopping work, will make it difficult to make the forwarding type of jamming effective. Forwarding type jamming, that is, after receiving the pulse signal modulation and then forwarded out, in order to achieve the effect of the jamming signal to cover the original signal, the speed of light is 300,000 km per second, 6.4us time light can go 1.92 km, so the repeater- jamming party - the receiver of the path length of this line minus the transmitter-receiver path length, and the difference cannot be greater than 1.92 km, i.e., the jamming machine must be within $1.92/2 = 0.96$ km, otherwise the repeater jamming will be completely ineffective, it is almost impossible to have a jammer so close and still not be found, so it is difficult to implement this kind of jamming.

3.7 Comparative Analysis of Several Different Jamming Styles

Table 1, depicting the comparison table of several different jamming styles. Targeted jamming boasts concentrated jamming power, but its implementation is exceedingly challenging, requiring not only prior knowledge but also robust jamming equipment performance, making it nearly impractical for unknown communication networks. Continuous spectrum blocking jamming, while easy to implement, suffers from low jamming utilization, resulting in inefficiency in certain scenarios. Partial-band blocking jamming, although relatively easier to implement, may exhibit limited effectiveness despite its concentrated jamming power. Conversely, comb spectrum blocking jamming, though somewhat complex to implement, offers high utilization and is a suitable method for jamming Ad Hoc signals compared to the former two methods. Despite the complexity associated with comb spectrum jamming, its concentrated power, high utilization rate, and absence of jamming distance restrictions make it an apt approach for disrupting Ad Hoc signals. However, repeater jamming necessitates close proximity due to elliptical restrictions.

Table 1. Comparison table of several different jamming styles.

Jamming techniques	Power utilization rate	Jamming distance	Difficulty level
Targeted jamming	High	Limitless	Easy
semi- targeted jamming	Average	Limitless	Easy
Continuous spectrum blocking jamming	Low	Limitless	Easy
Partial-band interdiction jamming	Average	Limitless	Easy
Comb-spectrum blocking jamming	High	Limitless	Average
Track-and-forward jamming	High	restrictive	Difficult

Given the inherent vulnerabilities of Ad Hoc networks to various jamming techniques, developing robust anti-jamming strategies is crucial. Effective measures include Frequency Hopping Spread Spectrum (FHSS), Direct Sequence Spread Spectrum (DSSS), adaptive power control, cognitive radio networks, advanced error correction techniques, and directional antennas. These techniques enhance the resilience of Ad Hoc networks by making communication more difficult to target, dynamically adjusting transmission parameters, utilizing less congested frequencies, and focusing transmissions directionally. Integrating these methods with cognitive electronic warfare capabilities leads to more sophisticated and adaptive defenses, ensuring robust and reliable communications even in contested environments.

4 Cognitive-Based Routing Protocol Techniques

Executing attacks against routing protocols requires familiarity with the network and the manipulation of information to deceive network nodes, achieving the attack's objectives through forgery and tampering [12]. However, obtaining such knowledge for a particular Ad Hoc network is challenging due to the relative scarcity of relevant information, making a priori data acquisition difficult.

In the realm of cognitive electronic warfare, reconnaissance aircraft autonomously navigate the electromagnetic battlefield, discerning threat radiation sources' states and inferring their electromagnetic behaviors [13]. Within this framework, the cognitive reconnaissance module serves as the cornerstone of cognitive communication electronic warfare, enabling environment perception, signal classification, and signal identification tasks. For Ad Hoc networks, at the physical signal level, data and control groups follow a specific sequence. This allows for capturing and parsing battlefield signals, conducting feature extraction and clustering, discerning signals directed to various nodes, and extracting relevant control groups to execute replay attacks against Ad Hoc routing protocols. Figure 7, depicting the flowchart of signal cognition and replay attack.

The CSMA/CA method introduced by IEEE 802.11 for Ad Hoc networking system is used to listen to the channel to avoid data transmission conflicts between sites [14].

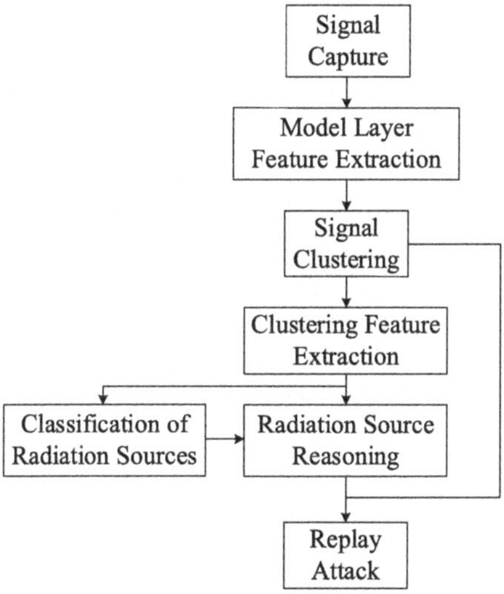

Fig. 7. Flowchart of signal cognition and replay attack.

When a node intends to transmit a packet exceeding a certain threshold size to another node, it first sends a small Request-To-Send packet (RTS) to the target node, indicating the duration of the upcoming communication. Upon receiving the RTS packet, the target node responds with a corresponding Clear-To-Send (CTS) packet, specifying the duration of channel occupancy for the communication.

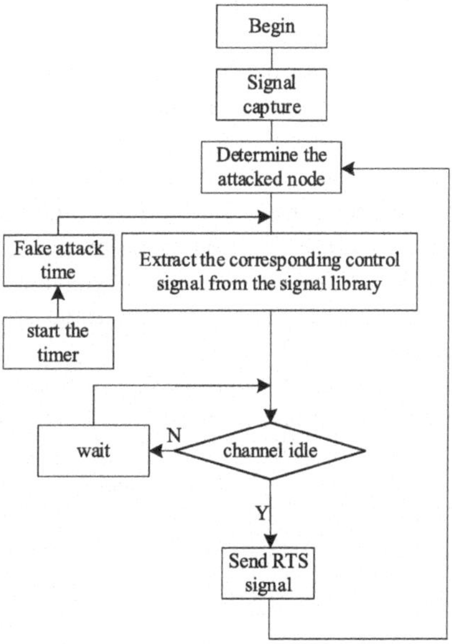

Fig. 8. Flowchart of channel occupancy attack.

The attack method targeting Ad Hoc MAC layer involves channel preemption, where nodes disregard MAC protocol rules by transmitting packets or control frames directly, regardless of the channel's state—whether idle or occupied. This action effectively monopolizes the channel, preventing other nodes from sending or receiving packets.

Figure 8, depicting the flowchart of channel occupancy attack. The attacking node intercepts and stores the enemy's designated node RTS/CTS packets, then retransmits them within the network. When any station receives this packet, it will refrain from attempting to access the channel for the specified duration. Consequently, the entire channel's availability is wasted due to the attacking node's deception. If the attacking node continually forges and broadcasts these RTS/CTS packets at a high frequency, it will monopolize the network channel within its range indefinitely. This results in the network segment being effectively paralyzed.

5 Conclusion

Through an in-depth analysis of the distinctive attributes of Ad Hoc networks, various techniques for jamming communication signals, and the inherent limitations of network protocols, it has become clear that the prevailing strategies for signal interference in Ad Hoc networks are a mixed bag of advantages and drawbacks. For example, while long-range jamming holds potential, it is considerably power-intensive. On the flip side, forward jamming can be highly efficient yet necessitates close proximity to the target. Leveraging advanced cognitive electronic warfare methods offers the possibility to precisely identify and extract specific signals within enemy Ad Hoc networks. This precision aids in executing targeted assaults on network protocols via specialized control frames, thereby facilitating communication disruption with reduced power expenditure.

Despite the limitations associated with traditional jamming and interference strategies, the employment of modern cognitive electronic warfare techniques heralds new avenues for deploying network countermeasures against selective protocols in unfamiliar Ad Hoc networks. This modern approach substantially bolsters our capability to incapacitate enemy communications with greater efficiency and efficacy. Moreover, by embracing these anti-jamming techniques, Ad Hoc networks can substantially bolster their resilience against interference efforts. Furthermore, the integration of such methods with cognitive electronic warfare capabilities paves the way for the development of more intricate and adaptable defense mechanisms, guaranteeing robust and reliable communication channels even in highly contested environments.

Acknowledgement. This work has received support from the National Natural Scien-ce Foundation of China under Grant 62171158 and National Key Laboratory of Air-based Information Perception and Fusion and Aeronautical Science Foundation of China under Grant No. 20230001004002. It was also supported by the Funda-mental Research Funds for the Central Universities under Grant 2242022k60006.

References

1. Sathyaprakash, B.P., Kotari, M.: Petal region analysis of improved petal ant routing for mobile ad hoc networks. SN Comput. Sci. **5**(5), 529 (2024)

2. Royer, E.M., Chai-Keong, T.: A review of current routing protocols for ad hoc mobile wireless networks. IEEE Pers. Commun. **6**(2), 46–55 (1999)
3. Niu, L., Wei, H., Jin, L.: Optimization design of networked control system with communication jamming. J. Phys. Conf. Ser. **1284**(1), 445–450 (2019)
4. Sharma, A.K., Kumar, R.: A Comprehensive survey of DDoS attacks: evolution, mitigation and emerging trend. In: 2024 3rd International Conference on Power Electronics and IoT Applications in Renewable Energy and its Control, PARC, pp. 185–188, IEEE, Mathura (2024)
5. Jie, Z., Gang, W., Xiaowei, Y.: Comparative analysis of sweeping frequency jamming and non-sweeping frequency jamming in partial-band jamming based on projectile-carried communication jamming. In: 2020 IEEE 20th International Conference on Communication Technology, ICCT, pp. 379–383. IEEE, Nanning (2020)
6. Nishanth, N., Mujeeb, A.: Modeling and Detection of flooding-based denial-of-service attack in wireless ad hoc network using bayesian inference. IEEE Syst. J. IEEE **15**(1), 17–26 (2021)
7. Khalil, M., Azer, M.A.: Sybil attack prevention through identity symmetric scheme in vehicular ad-hoc networks. In: 2018 Wireless Days, WD, pp. 184–186. IEEE, Dubai (2018)
8. Zhitao, Y., Hang, H., Dan, W.: Advanced optimization for subarray structure in electronic warfare environment. In: 2013 IEEE International Conference on Microwave Technology & Computational Electromagnetic*s*, pp. 30–34. IEEE, Qingdao (2013)
9. Liu, S., Ma, L., Hu, X.: The optimizing of LDPC coded frequency-hopping communication over blocking jamming. In: 2018 International Conference on Electronics Technology, ICET, pp. 354–360. IEEE, Chengdu (2018)
10. Qian, L., Zhang, J., Chen, D.: Simulation Method of Jamming Performance of Projectile-Carried Communication Jammer Based on OPNET Platform. In: Lecture notes in electrical engineering, vol. 576, pp. 445–451. Springer, Singapore (2019)
11. Jinyang, H., Ziyang, C., Zishu, H.: Joint design of detecting and screening waveforms against interrupted sampling repeater jamming with inaccurate jamming knowledge. Digit. Signal Process **130**, 103679 (2022)
12. Stępień, K., Poniszewska-Marańda, A.: Security methods against black hole attacks in vehicular Ad-Hoc network. In: 2020 IEEE 19th International Symposium on Network Computing and Applications, NCA, pp. 1-4. IEEE, Cambridge (2020).
13. Zhou, H.: An introduction of cognitive electronic warfare system. In: Lecture Notes in Electrical Engineering, vol. 517, pp. 1202–1210. Springer, Singapore (2019)
14. Amuru, S., Buehrer, R.M.: Optimal jamming using delayed learning. In: 2014 IEEE Military Communications Conference, pp. 1528–1533. IEEE, Baltimore (2014)

Base Station Location Optimization Framework Based on Digital Twins for 6G Networks

Zongxuan Sha[1], Ru Huo[1,2(✉)], and Tao Huang[2,3]

[1] School of Information and Science Technology, Beijing University of Technology, Beijing 100124, China
huoru@bjut.edu.cn

[2] Purple Mountain Laboratories, Nanjing 211100, China

[3] State Key Laboratory of Networking and Switching Technology, Beijing University of Posts and Telecommunications, Beijing 100876, China

Abstract. With the rapid development of areas such as the Internet of Things (IoT) and the Internet of vehicles (IoV), the surge in traffic demand could cause network congestion and interfere with infrastructure planning. With the increasing popularity of 6G networks, wider network bandwidth also means smaller coverage of base stations (BSs). Therefore, an efficient wireless BS deployment method is urgently needed. This paper proposes a BS location optimization framework, trying to solve the problem of wireless BS location from a micro point of view, the problem is divided into traffic demand prediction and BS location optimization. Specifically, the framework is divided into physical layer, virtual layer and control layer. The physical layer contains all physical objects. In the virtual layer, we use real-time profiles based on digital twins (DT) to generate traffic. This is followed by incorporating the most important traffic logs into the forecast. In the control layer, we use the transformer architecture to predict long period network traffic and generate the traffic heat map. On the basis of forecast traffic, we design a BS location optimization module based on deep deterministic policy gradient (DDPG), and output a forward-looking BS location scheme. The experimental results show that the framework and algorithm designed in this paper can effectively improve the traffic coverage of BSs.

Keywords: 6G Networks · Base Station Location Optimization · Deep Reinforcement Learning · Deep Deterministic Policy Gradient · Digital Twins

1 Introduction

1.1 A Subsection Sample

In recent years, with the rapid development of mobile communication technologies, network service deployment and the increasing popularity of portable handheld devices, network traffic has increased rapidly [1]. According to Cisco's report of global mobile data traffic, global mobile data traffic has grown sevenfold over the past few years [2]. As

X. Li et al. (Eds.): 6GN 2024, LNICST 615, pp. 14–27, 2025.
https://doi.org/10.1007/978-3-031-93193-2_2

a result, faster 6th generation mobile networks (6G) are applied [3]. Internet of Things (IoT), which supports large-scale devices connectivity and the internet of everything, is one of the main goals of 6G [4]. Cellular network is still retained as the basis for mobile communication networks. Therefore, wireless network has become an indispensable part of people's daily production and life. As a key component of wireless network, the location selection of BS s has an important impact on network performance and user experience [5].

In 6G networks, continuous network parameter optimization helps improve network performance and user experience. Reasonable BS location planning before dense network deployment could reduce network deployment costs and improve network coverage and capacity [6]. The traditional wireless BS location method often fails to fully consider the overall optimization of the network, and the prediction in the aspects of signal coverage and network capacity may not be accurate enough and lack of predictability [7, 8]. Therefore, it is of great practical significance to study how to optimize the location of BSs.

To solve the above problems, we propose a BS location optimization framework (BSLOF) based on DT. From the perspective of structure, the proposed framework is divided into physical layer, virtual layer and control layer. The physical layer contains all physical objects. In the virtual layer, the mapping from physical objects to digital space is established by constructing DT objects. The historical data in the DT is used to train the model and verify the BS location strategy. The control layer includes traffic prediction module and BS location optimization module. In traffic demand forecasting, in our proposed architecture, a DT is constructed for each physical network element [9]. We use transformer architecture to predict future data based on the network's historical traffic data. The transformer architecture has excellent long sequence sample prediction performance [10]. In the part of BS location optimization, considering the correlation between different BS locations, we built an optimization algorithm based on deep reinforcement learning (DRL) to determine the optimal location of the BS according to the predicted traffic demand and BS performance.

This paper has the following contributions:

1. The optimization framework of BS location based on DT is designed. The framework is divided into physical layer, virtual layer and control layer. We construct DT s of the physical objects and use the DT to collect historical traffic. The architecture includes traffic prediction module and BS location optimization module to realize the function of optimizing BS location according to future traffic.
2. A traffic prediction module based on transformer is designed, and the prediction model is trained according to historical traffic. The model could output long-period forecast traffic according to different regional types and form a heat map of forecast flow. The heat map could directly reflect the different traffic characteristics of different types of areas, and provide basic data for the subsequent location of BSs.
3. The BS location optimization module based on DDPG is designed. On the basis of the future traffic heat map, a DRL model is constructed to output the BS location strategy. To achieve the purpose of improving the traffic coverage of BS.
4. Performing data cleaning on the SDNLib dataset to reduce the influence of default data and high jitter data on traffic prediction and base location optimization algorithm.

The results show that the proposed algorithm outperforms the other methods in terms of training time, prediction accuracy and flow coverage.

2 Background Information and Related Work

2.1 Background Information

In this section, we will briefly introduce the methodologies employed in this paper. This includes providing background information on three key approaches: DDPG, the transformer architecture and DT.

Deep Deterministic Policy Gradient
Reinforcement Learning (RL) is an approach to machine learning that focuses on making an agent learn how to act in an unknown environment to maximize cumulative rewards [11]. Unlike supervised learning, RL does not require pre-labeled data, but instead lets agents learn how to act by interacting with the environment and based on the results of those interactions (i.e., rewards) [12].

DDPG is a DRL algorithm mainly used to solve the problem of continuous action space, and its architecture is shown in Fig. 1.

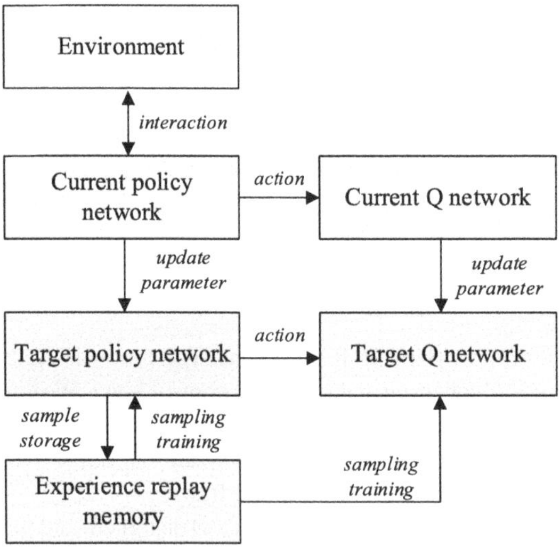

Fig. 1. Deep deterministic policy gradient algorithm architecture.

DDPG is a RL algorithm that combines Actor-Critic framework and deep neural networks to optimize the choice of continuous actions by learning a deterministic strategy [13–15]. Among them, the policy network is used to output the action, that is, to select the best action based on the current state. The Q network is used to evaluate the value of the action output by the policy network, that is, to evaluate the expected reward for a given state and action. DDPG uses a deterministic strategy to output the values of successive actions directly, rather than the probability distribution of the output actions. This makes the strategy easier to optimize and stabilize. The experience replay mechanism could reduce correlations between samples and allow the agent to reuse previous experiences.

Digital Twin

The core idea of DT is to complete mapping in virtual space by integrating physical model, sensor update, operation history and other data, so as to reflect the whole life cycle process of the corresponding physical equipment [16]. Driven by data, DT could realize functions such as monitoring, simulation and prediction. Due to the seamless integration of digital space and physical space, the importance of DT is increasingly recognized in academia and industry. So far, DT has been successfully applied in different industries [17].

Transformer Architecture

Transformer architecture is a neural network architecture based on the self-attention mechanism, which is originally used in the field of natural language processing (NLP) and has become the basis for many modern NLP models, such as BERT, GPT [18, 19]. Compared to recurrent neural network (RNN) and long short-term memory (LSTM), the success of Transformer architecture is due to its effective self-attention mechanism, multi-head attention mechanism and deep network structure, which enables it to efficiently process long sequences and capture complex dependencies.

2.2 Related Work

There are some works on the selection strategy of BS locations. In the following, we overview some relevant work.

Reference [20] proposes an optimal location method of BS based on propagation path models. The authors analyze that if the user set is symmetric, the optimal BS position is independent of the path loss index. The correctness and effectiveness of the algorithm are verified by numerical experiments.

In reference [21], the authors analyze that BS is a central hub for data collection and communication, and their location has a great impact on the performance and energy efficiency of the entire network. The authors propose a meta-heuristic based optimization location allocation algorithm for BSs, aiming to find a compromise to appropriately place BS in the center of the deployment area and outside.

Wang et al. argue in reference [6] that some studies use hop count to measure the transmission delay, while ignoring the actual transmission process. The authors first establish a network reliability model and propose a new model of a network design problem with relays to optimize the location of BSs.

Reference [22] uses K-means algorithm, analog degradation algorithm, linear programming and other methods to establish a model and find out the location of BS construction with the lowest construction cost. The proposed method could achieve a higher network coverage at a lower construction cost.

Reference [23] believes that with the development of the network, the wider the bandwidth, the more limited the coverage of the BS. Weak network coverage areas may occur. The author establishes a multi-objective programming model and uses a heuristic algorithm to solve it.

We find that the traditional BS location optimization method is based on the current environmental output scheme and lacks consideration of future traffic changes. Some optimization methods are not robust to different environments. Therefore, this paper studies the BS location optimization framework based on DT, uses the historical data of DTs to train the prediction model, and uses DRL to output the BS location strategy.

3 Base Station Location Optimization Framework

In BSLOF, there are three layers, namely physical layer, virtual layer and control layer, as shown in Fig. 2.

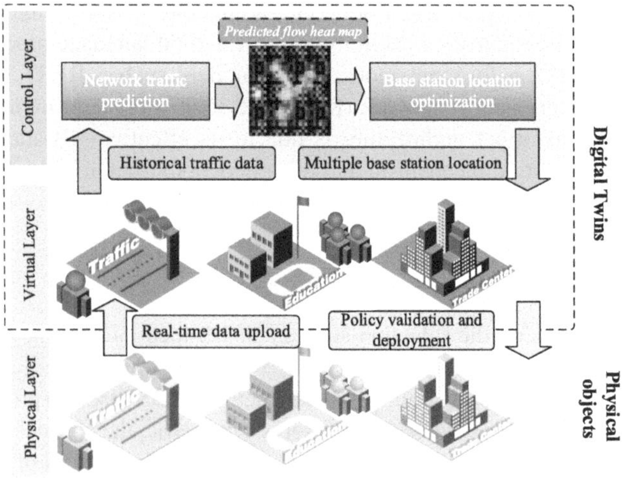

Fig. 2. BS location optimization framework based on DTs.

3.1 Physical Layer

The physical layer contains all physical network elements, which are physical devices and objects on the actual network. In urban scenarios that require BS optimization, we also consider a unit containing a large amount of communication equipment as a physical network element. For example, the trading center contains a wide variety of communication devices, and the traffic characteristics of different devices are different, but if it is regarded as a whole, it also shows periodic traffic characteristics. By abstracting multiple network devices in the same location into one physical network element, the granularity of system observation can be reduced.

In addition, physical network elements carry actual services and generate traffic data. This data supports the functionality of other layers.

3.2 Virtual Layer

In BSLOF, we build DTs for physical network elements. These DTs form the virtual layer. A DT is a virtualized representation of a physical entity. On the one hand, DTs can obtain data of physical entities in real time for monitoring and data analysis. Due to the rapid growth of network traffic, BSLOF utilizes the predicted traffic data to optimize the location of BSs, so that the optimization scheme can meet the network demand for a period of time in the future and overcome the problem of insufficient prediction of traditional methods.

On the other hand, real business can also be run in a DT to simulate the effects of deploying the business in a real-world environment. In the traditional BS location optimization method, the effect of the planning scheme can be verified only after the actual network deployment. Therefore, BSLOF uses the virtual layer to simulate different schemes, which helps to reduce the simulation cost.

3.3 Control Layer

The control layer mainly consists of two modules, the flow prediction module based on transformer and the BS location optimization module based on DRL.

The transformer-based traffic prediction module uses the historical traffic of the DT collected from the virtual layer to generate the predicted traffic. After the predicted traffic is obtained, the control layer then obtains the flow heat map according to the position of the DT. The heat map describes the future flow demand of different positions in the region through the form of a two-dimensional matrix.

After obtaining the traffic heat map, we designed the BS location optimization model based on DRL. For the problem of BS position optimization, the role of BSs will affect each other, that is, the change of the position of one BS will affect the effect of other BSs. This problem could be solved by DRL algorithm. For general RL, agents interact with the environment to update their own strategies guided by reward signals, with the goal of maximizing the expectation of future rewards, but only one agent interacts with the environment. DRL is an extended form of RL, which aims to study how to learn and make decisions in an environment where the agent interact and collaborate.

4 Traffic Prediction Module

This model consists of a transformer block in series. The structure of the prediction model is shown in Fig. 3.

The raw long-period traffic data is passed to the transformer block to generate to generate forecasts of future traffic. The traffic prediction module consists of multiple transformer blocks stacked sequentially. Each transformer block consists of an attention part and a feed forward part.

In the attention part, multi-head attention means that the attention mechanism is evaluated multiple times, and the *Query*, *Key*, and *Value* parameters are divided into M parts, and then each part is processed by an independent attention head.

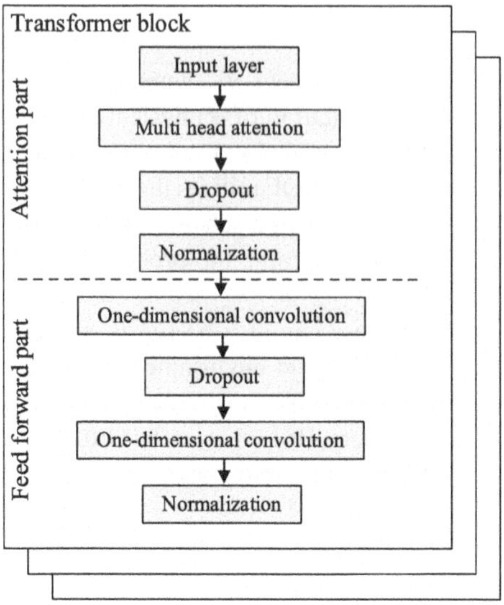

Fig. 3. BS location optimization framework based on DTs.

These functions are as follows:

$$Q = XW_q \tag{1}$$

$$K = XW_k \tag{2}$$

$$V = XW_v \tag{3}$$

$$M_A = \frac{QK^T}{\sqrt{D_K}} \tag{4}$$

$$H = softmax(M_A)V = W_A V \tag{5}$$

where, the W_q, W_k, and W_v are parameters matrix, and D_K represents the dimensionality of a query vector in the queries matrix K. The attention matrix M_A is obtained from the attention scoring function, and then the attention weight matrix W_A is obtained through the *softmax* function. Multi-head attention is a linear transformation that combines the results of multiple attention mechanisms. The method is to transform the Q, K, and V vectors into several sets of $Q/K/V$ matrices using distinct linear projection matrices. After completing these transformations, the attention calculations are performed on these matrices. The output value is obtained by multiplying with the parameter matrix W^O.

Assuming h projection spaces, the multi head attention is calculated as:

$$Multihead(H) = W^O[h_1; h_2, \ldots, h_h] \tag{6}$$

The dropout layer randomly drops some neurons during training so that the network cannot rely on specific neurons, reducing model overfitting and improving generalization ability. The normalization layer uses normalization data to speed up model training and stabilize the training process.

In the feed forward part, the one-dimensional convolution layer, dropout layer, and normalization are mainly included. One-dimensional convolution layer could effectively extract local features of historical traffic data. The Dropout layer and normalization function similarly to the attention part.

A complete traffic prediction module is stacked with multiple transformer blocks, which takes historical traffic data as input and outputs the forecast traffic for a certain period in the future. Due to the attention mechanism in transformer, the architecture can handle long time series data better than traditional RNN and LSTM. Therefore, this module is built on the transformer architecture to forecast network traffic over long-time spans.

5 Base Station Location Optimization Module

In this section, we will introduce the BS location optimization module. Based on DDPG, the module takes the traffic prediction information and BS location information obtained in Sect. 4 as module input, and outputs the BS location strategy. The model structure is shown in the Fig. 4.

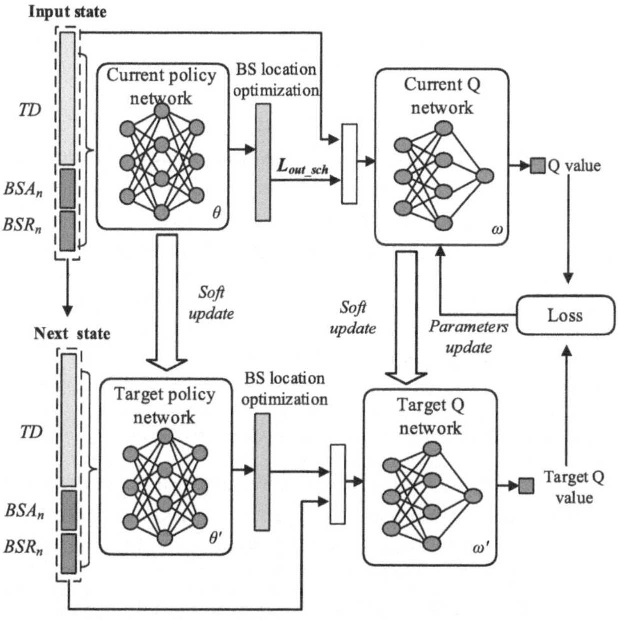

Fig. 4. Structure of BS location optimization model.

The module consists of four neural networks, namely current policy network, current Q network, target policy network and target Q network.

Current policy network takes traffic distribution and BS status as input, and outputs the BS location optimization strategy. Current Q network evaluates the policy output by the current policy network and outputs the corresponding Q value. Target policy network and target Q network are the key components used to stabilize the training process. The two target networks have the same structure as the current networks. By introducing the target network, the target value could be kept relatively stable over a period of time, thus reducing the instability of training. In addition, the problem of gradient explosion and gradient disappearance of the current network could be effectively alleviated through the soft update mechanism, so that the gradient change in the training process is more stable.

State is the input of the BS location optimization module. Set the number of BSs to n and the length and width of the flow heat map to m, then the state vector can be expressed as $s_t = [TD, BSA_n, BSR_n]$. TD is to reshape the traffic heat map into a one-dimensional vector, representing the discrete traffic distribution value, and its length is 2 m. BSA_n represents the capacity of the BS to handle traffic. BSR_n indicates the coverage radius of the BS. Both BSA_n and BSR_n have length n. The overall length of the state vector is 2m + 2n.

Action is the location policy output by the model, $a_t = [L_{out\_sch}]$. This vector contains the location information of all BSs. Since this paper considers deploying BSs in a two-dimensional environment, the length of this vector is 2n.

Reward response outputs the quality of the location selection strategy. This paper takes the traffic coverage rate as the optimization objective, as shown in formula 7.

$$r_t = traffic_{BS}/traffic_{total} \tag{7}$$

where, $traffic_{BS}$ indicates the traffic demand that the BS can meet. $traffic_{total}$ indicates the total traffic demand in the network. This value reflects the degree to which the BS location strategy can meet the traffic demand, and also provides the optimization direction for the model parameters. The BS location optimization model trained with this reward function can continuously optimize the BS location to obtain higher traffic coverage.

6 Performance Evaluation

6.1 Experimental Settings

In this section, we present performance evaluation results for the proposed BSLOF. Traffic prediction module and BS location optimization module are constructed by Keras, a high-level API based on Tensorflow and Pytorch. The hardware configuration and software versions are shown in Table 1.

We obtained the traffic data from SNDLib and simulated the service data in the physical layer. The data set is based on real data from European telecom operators. Meanwhile, we structured a DT object with the same topology as the physical network and to capture historical traffic data. The traffic heat map we constructed is a 2-dimensional image with

130 discrete values in length and width. Set the coverage radius of a single BS to 30, and the coverage of the BS could overlap.

The input vector length of the traffic prediction module is 144 and the output vector length is 10. At the same time, the traffic data is discretized, and the average traffic is taken with a span of one minute. After data processing, 1440 discrete traffic values could be generated for one day's traffic flow. Each transformer block contains 4 heads with a head size of 256. The traffic prediction module consists of four transformer blocks in series. To prevent overfitting, the proposed model is set to dropout ratio of 0.25.

Table 1. Hardware configuration and software version.

Items	Stats
OS	Ubuntu 22
CPU	12th Gen Intel(R) Core(TM) i7-12700H
GPU	NVIDIA GeForce RTX 3050
Memory	32.0 GB
Python	Ver. 3.12
Tensorflow	Ver. 2.16
Keras	Ver. 3.4.0

In BS location optimization module, the policy networks and Q networks contain three hidden layers, which contain 256, 512 and 256 neurons respectively. The activation function of the hidden layer is set to ReLU. This activation function introduces nonlinear transformation into the model, and is easy to calculate the gradient, which could reduce the training time of the model. The learning rate is set to 1e−4. Discount rate γ is set to 0.99.

6.2 Performance Analysis of Traffic Prediction Module

In this part, we compare the convergence values of BSLOF proposed in this paper, LSTM, Bidirectional LSTM and RNN algorithms. The experimental results are shown in Fig. 5.

Since the traffic prediction module of BSLOF is based on transformer. Transformer utilizes an attention mechanism that allows you to process all elements of a sequence in parallel while training, instead of processing data step by step. This allows BSLOF to improve the training efficiency of the traffic prediction module. As can be seen from Fig. 5, the loss value of the BSLOF decreases more rapidly and reaches a stable state after 134 epochs. At this time, the other three algorithms are still in the middle stage of model training.

After model training, the prediction accuracy of the four prediction algorithms is shown in Table 2. We used Mean Absolute Error (MAE), Normalized Root Mean Square Error ($NRMSE$), and Regression Score (R^2) to compare performance.

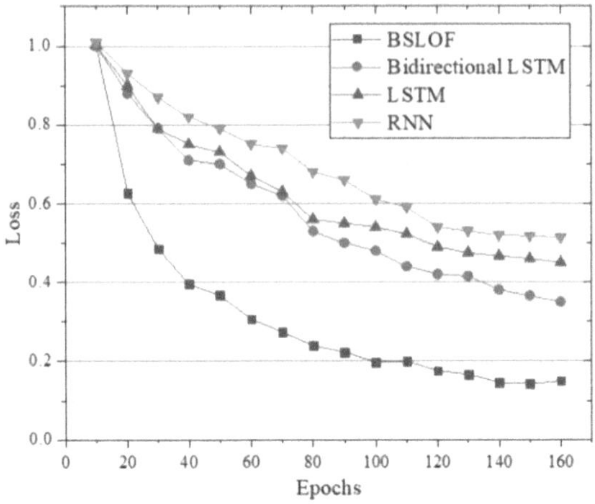

Fig. 5. Comparison of convergence of algorithms.

Table 2. Prediction accuracy of different algorithms.

Algorithm	*MAE*	*NRMSE*	R^2
BSLOF	**10.0521**	**0.6213**	**0.8854**
Bidirectional LSTM	13.0163	0.7519	0.7745
LSTM	14.7812	0.8218	0.7425
RNN	15.4342	0.8362	0.6812

For the above indicators, the smaller the value of *MAE* and *NRMSR*, or the closer the value of *R2* is to 1, the better the prediction performance of the algorithm. It can be seen from Table 2 that BSLOF achieves optimal performance in all three indicators. Compared with traditional baseline models, BSLOF has higher prediction accuracy for long time series data and faster model training time.

6.3 Performance Analysis of Base Station Location Optimization Module

After getting a more accurate forecast traffic, we built a forecast-based traffic heat map in the DT. In this section, we compare the performance of four algorithms for BS location optimization. The corresponding algorithms are Uniform location, PPO, BSLOF without forecasting and BSLOF. Uniform location is a location selection method that adopts uniform coverage of BSs in order and does not overlap the coverage range. PPO is a classical RL algorithm. In addition, on the basis of BSLOF, we also verify the optimization performance of BSLOF w-o forecasting according to the current traffic condition, without the traffic prediction module.

We gradually increased the number of BSs in the same size experiment scene to verify the traffic coverage that can be obtained by the BS location method output by different algorithms. The experimental results are shown in the Fig. 6.

As can be seen from Fig. 6, each algorithm will improve the traffic coverage as the number of BSs increases. Because BSs are uniformly deployed in the scene, the Uniform Location algorithm could not meet the requirements of traffic hotspots, so the traffic coverage rate is at the lowest level. The other three intelligent algorithms could adjust the location of the BS according to the traffic status. The PPO and BSLOF w-o forecasting algorithms output the BS location strategy according to the current traffic status, and the traffic coverage rate stays at about 65% with the increase of the number of BSs. BSLOF is based on the forecast traffic, and then carry out the BS location optimization, which can make the BS better cover the traffic hotspot area, and further improve the traffic coverage. In the end, BSLOF's traffic coverage stabilized at 83%.

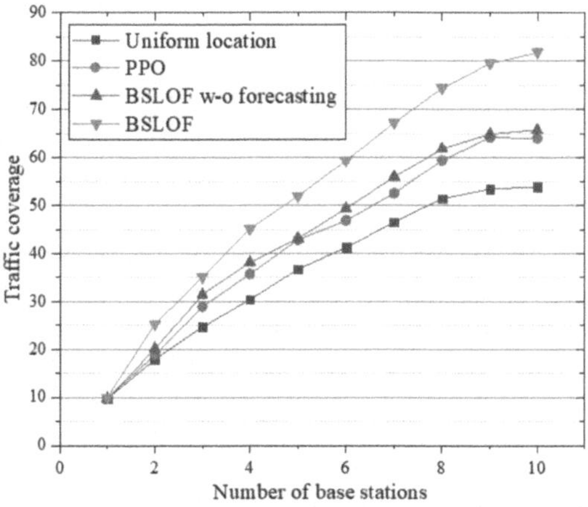

Fig. 6. Comparison of network coverage of different algorithms.

In order to further compare the optimization effects of different algorithms for BS location, we verify the value of traffic coverage that can be improved each time a BS is added. The result is shown in Fig. 7.

As can be seen from the above experimental results, except for the other three algorithms of BSLOF, with the increase of the number of BSs, the effect on traffic coverage all showed a rapid decline. Before increasing the number of BSs to 8, BSLOF maintained an average coverage increase of more than 8% each time the number of BSs is increased. This result also shows the excellent performance of BSLOF for BS location optimization. When the number of BSs increases to more than 8, in this experiment scenario, the network capability provided by BSs tends to be saturated, so the added value of traffic coverage of all algorithms shows a downward trend.

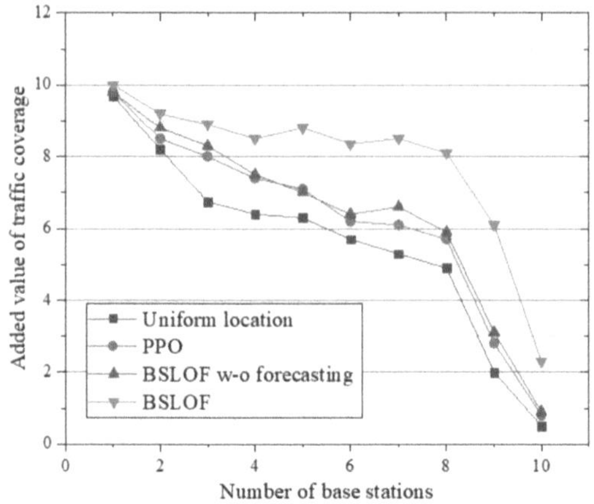

Fig. 7. Comparison of added value of traffic coverage.

7 Conclusion

With the development of 6G networks, the wider the bandwidth, the smaller the coverage of the BS. BS location strategy is more and more important to ensure network performance. Considering the long-term change of network traffic, this paper proposes a BS location optimization framework based on DT. The framework consists of physical layer, virtual layer and control layer, in which the virtual layer and control layer are in the DT object. Using the historical traffic data in DT, this paper first proposes a traffic prediction module based on transformer. The BS location optimization module based on DDPG is designed according to the predicted traffic. The DT also provides the function of verifying the location strategy of the BS. The experimental results show that compared with other baseline models, the BSLOF proposed in this paper could effectively improve the accuracy of traffic prediction and shorten the training time of the prediction model, and the BS location optimization module could also improve the traffic coverage. In the future, we plan to introduce other features such as user mobility and conduct experiments on larger-scale real-world data to verify the effectiveness and generalization ability of our framework.

Acknowledgement. This work was supported by the Research and Development Program in Key Areas of Guangdong Province (No. 2018B010113001).

References

1. Bennis, M., Debbah, M., Poor, H.V.: Ultrareliable and low-latency wireless communication: tail, risk, and scale. Proc. IEEE **106**(10), 1834–1853 (2018)
2. Index V N. Cisco Visual Networking Index: Forecast and Methodology, 2010–2015 (2011)
3. Mahboob, S., Liu, L.: Revolutionizing future connectivity: a contemporary survey on AI-empowered satellite-based non-terrestrial networks in 6G. IEEE Commun. Surv. Tutor. **26**(2), 1279–1321 (2024)

4. Chen, W., Liu, D., Bai, Y.: B5G/6G URLLC latency reduction method for multi-sensor industrial internet of things. IEEE Internet Things J. **11**(7), 11444–11459 (2023)
5. Stusek, M., Moltchanov, D., Masek, P., et al.: LPWAN coverage assessment planning without explicit knowledge of base station locations. IEEE Internet Things J. **9**(6), 4031–4050 (2021)
6. Wang, N., Xiao, Y., Tian, T., et al.: The optimal 5G base station location of the wireless sensor network considering timely reliability. Reliab. Eng. Syst. Saf. **236**, 109310 (2023)
7. Valavanis, I.K., Athanasiadou, G., Zarbouti, D., et al.: Base-station location optimization for LTE systems with genetic algorithms. In: European Wireless 2014; 20th European Wireless Conference. VDE, pp. 1–6 (2024)
8. Liu, Y., Huangfu, W., Zhang, H., et al.: An efficient geometry-induced genetic algorithm for base station placement in cellular networks. IEEE Access **7**, 108604–108616 (2019)
9. Jones, D., Snider, C., Nassehi, A., et al.: Characterising the Digital Twin: a systematic literature review. CIRP J. Manuf. Sci. Technol. **29**, 36–52 (2020)
10. Han, K., Wang, Y., Chen, H., et al.: A survey on vision transformer. IEEE Trans. Pattern Anal. Mach. Intell. **45**(1), 87–110 (2022)
11. Moerland, T.M., Broekens, J., Plaat, A., et al.: Model-based reinforcement learning: a survey. Found. Trends® Mach. Learn.**16**(1), 1–118 (2023)
12. Kaelbling, L.P., Littman, M.L., Moore, A.W.: Reinforcement learning: a survey. J. Artif. Intell. Res. **4**, 237–285 (1996)
13. Wang, D., Hu, M.: Deep deterministic policy gradient with compatible critic network. IEEE Trans. Neural Netw. Learn. Syst. **34**(8), 4332–4344 (2021)
14. You, S., Diao, M., Gao, L., et al.: Target tracking strategy using deep deterministic policy gradient. Appl. Soft Comput. **95**, 106490 (2020)
15. Ning, L., Zhou, M., Hou, Z., et al.: Deep deterministic policy gradient for high-speed train trajectory optimization. IEEE Trans. Intell. Transp. Syst. **23**(8), 11562–11574 (2021)
16. Tao, F., Xiao, B., Qi, Q., et al.: Digital twin modeling. J. Manuf. Syst. **64**, 372–389 (2022)
17. Tao, F., Zhang, H., Liu, A., et al.: Digital twin in industry: state-of-the-art. IEEE Trans. Ind. Inf. **15**(4), 2405–2415 (2018)
18. Tay, Y., Dehghani, M., Bahri, D., Metzler, D.: Efficient transformers: a survey. ACM Comput. Surv. **55**(6), 1–28 (2022)
19. Khan, S., Naseer, M., Hayat, M., et al.: Transformers in vision: a survey. ACM Comput. Surv. (CSUR) **54**(10s), 1–41 (2022)
20. Kalantari, E., Loyka, S., Yanikomeroglu, H., et al.: Optimal location of cellular base station via convex optimization. In: 2020 IEEE International Black Sea Conference on Communications and Networking (BlackSeaCom), pp.1–6 (2020)
21. Chaurasia, S., Kumar, K.: MBASE: meta-heuristic Based optimized location allocation algorithm for baSE station in IoT assist wireless sensor networks. Multimed. Tools Appl. 1–33 (2023)
22. Li, Z.: Research on 5G base station location problem based on simulated annealing algorithm. In:Proceedings of the 7th International Conference on Intelligent Information Processing, pp. 1–5 (2022)
23. An, Y., Li, C., Duan, W., et al.: Base station location determination model based on 5G network coverage. In: 2022 IEEE 5th International Conference on Information Systems and Computer Aided Education (ICISCAE), pp. 736–740. IEEE (2022)

Modeling and Optimization of Random Access in NB-IoT Network with Heterogeneous Users

Haiheng Ye and Peiran Wu(✉)

Sun Yat-sen University, Guangzhou, China
yehh5@mail2.sysu.edu.cn, wupr3@mail.sysu.edu.cn

Abstract. Narrowband Internet of Things (NB-IoT) is a fundamental technology for massive Machine Type Communications (mMTC) usage scenarios. Before uplink communication, a user equipment (UE) needs to establish connection to the base station by performing a random access (RA) procedure. In this paper, focusing on the contention-based random access of a heterogeneous NB-IoT network, we provide a model based on Markov chains to obtain the steady-state distribution of each UE type in the network, and then further evaluate the system performance. Based on the analysis of the model, we adopt an optimization strategy that configures parameters for each UE type individually. Considering the fairness of the system, we propose an optimization algorithm based on simulated annealing. Simulation results show that the proposed model is valid, and the proposed optimization algorithm can improve the overall throughput while taking count of the fairness among users with an acceptable complexity.

Keywords: Narrowband Internet of Things (NB-IoT) · contention-based random access · heterogeneous network

1 Introduction

Internet of Things (IoT) is a concept described as the automatic connection and interaction among various devices via Internet and communication technology without human intervention. IoT enables a mass of promising and intelligence technologies, which will change our lives in the following future.

Narrowband Internet-of-Things (NB-IoT), introduced by the 3rd Generation Partnership Project (3GPP), is a Low-Power Wide-Area Network (LPWAN) technology designed for the massive Machine-Type Communications (mMTC) usage scenario [1]. NB-IoT applies to scenarios such as smart metering, smart logistics and environment monitoring. In these scenarios, a great number of connected devices are deployed, which typically transmit a small volume of data with a high tolerance for time delay. NB-IoT inherits the framework of 4G Long Term Evolution (LTE) while only requiring 180 kHz bandwidth. NB-IoT has

X. Li et al. (Eds.): 6GN 2024, LNICST 615, pp. 28–44, 2025.
https://doi.org/10.1007/978-3-031-93193-2_3

distinctive technical features [2] including coverage enhancement, high throughput, insensitivity to delay, high capacity for massive devices, low device cost and low energy consumption.

Before uplink communication, a user equipment (UE) needs to conduct a random access (RA) procedure so as to establish connection to the base station. Similar to LTE, the contention-based RA procedure in NB-IoT consists of four steps [3]. Firstly, every accessing UE selects a RA preamble randomly and launches it to the base station. Secondly, the base station replies the UE with a corresponding RA response after receiving the preamble successfully. Thirdly, the UE launches a connection request to the base station. Lastly, the base station replies the UE with message containing a contention resolution identity.

NB-IoT adopts a single-tone waveform with frequency-hopping as RA preamble [4]. The RA preamble is made of symbol groups that consist of one cyclic prefix (CP) and five identical single-tone OFDM symbols. Every symbol group occupies a specific subcarrier according to the frequency hopping mechanism. If there are two or more UEs transmitting the same RA preamble at the same time, a preamble collision occurs which makes these collided UEs fail in this RA attempt.

In contention-based RA procedure, a series of RA congestion solutions are introduced [5], such as access class barring (ACB) scheme and the back-off scheme. In ACB scheme [6], before transmitting a RA preamble, the UE needs to conduct an ACB judgment according to a probability called ACB factor. If passing the judgment, the UE continues with the RA procedure. Otherwise, the UE should wait for some time and then perform a new ACB judgment. In the back-off scheme [7], if the UE fails in a RA attempt, it will wait for some time and then retransmits a RA preamble until the maximum number of RA attempts is reached.

1.1 Related Works

In recent years, there has been a lot of researches on the modeling and optimization of contention-based RA procedure in NB-IoT. For the situation that massive access bursts in a short time, authors in [8] model the RA procedure in the form of a multi-band multi-channel slotted ALOHA protocol when considering the coverage enhancement (CE) level mechanism and the back-off scheme. They evaluate the access success probability and the average access delay based on the model, and then propose a joint optimization algorithm for them. Based on the analysis of model in [8], the authors carry on their research in [9,10]. In [9], they consider the fairness of the normalized throughput among CE levels, and then propose a strategy of parameter configuration to maximize the overall throughput at the prerequisite of an acceptable fairness level. In [10], they put forward an idea that the back-off scheme in the MAC layer can play a role in time diversity effect for preamble detection in the physical layer. Then they propose an optimization strategy to improve the resource efficiency by making a trade off between preamble repetition in the physical layer and preamble retransmission in the MAC layer. For the steady state of RA procedure in a NB-IoT network, authors in [11] evaluate the success rate of a single RA attempt according to SINR when considering the spatial distribution of UEs and base stations. Then they further evaluate the access success probability of the system in steady state by considering RA congestion solutions and traffic models. In the perspective of steady state distribution when considering the back-off scheme and buffer queue, authors in [12] propose a model based on Markov chains to analyze the throughput of a NB-IoT network. Similarly, authors in [13] propose a model in the form of Markov chain to evaluate the throughput when considering the back-off scheme and CE levels. Further more, they provide the analytical expression of the state distribution.

In summary, it is important to model and evaluate the performance of RA procedure in an analytical form, which can contribute to the deployment and real-time optimization of a NB-IoT network. To simplify the analysis, the existing researches mostly assume that all UEs deployed in the network have the same statistic characteristics, such as traffic model, channel quality and so on. However, in mMTC usage scenario, an IoT system usually involves a variety of function modules so as to be "smart" enough to support a complex and comprehensive application. The UEs in different function modules may be heterogeneous in design, traffic flow and wireless environment. For example, as a typical application in mMTC, a smart energy grid [14] needs to deploy various UEs, including sensors for residential electricity monitoring, sensors for industrial electricity monitoring, sensors for power transmission monitoring, controllers for load managements and so on. Obviously, taking account of the characteristics of heterogeneous UE types can make the model more practical and comprehensive.

1.2 Contribution

In this paper, we model and optimize the RA procedure in a NB-IoT network for a heterogeneous network case, where several UE types heterogeneous in traffic model and channel quality are deployed. Our contributions in this paper are summarized as follows:

- An analytical model based on Markov chains is designed for a heterogeneous network. In the form of Markov chain, we define the access states of each UE type by considering buffer queue, ACB scheme and the back-off scheme. The Markov chain will reach a steady state after a few iterations, and then we evaluate the performance metrics of each UE type.
- An optimization algorithm based on simulated annealing is proposed, which configures parameters for each UE type individually. This algorithm makes a trade off between the overall throughput and the fairness among UE types.

The rest of this paper is organized as follows. In Sect. 2, we introduce the system model of RA procedure in NB-IoT. In Sect. 3, we introduce the analytical model of RA procedure based on Markov chains. In Sect. 4, we present an optimization algorithm based on simulated annealing. In Sect. 5, we discuss the results of simulation.

2 System Model

In a heterogeneous network, we assume that there are totally N UEs deployed in the system, and they can be divided into N_{class} UE types. Let the number of UEs with UE type i be denoted as N_i, and these UEs with the same UE type have the same statistic characteristics and configuration.

2.1 Traffic Model

For mMTC usage scenario, the traffic model is subject to a Beta distribution when massive access bursts in a short time, and the traffic model is subject to a uniform distribution or a Poisson process during a long-term observation [15]. Hence, it is reasonable to assume that the traffic models of UE types are subject to Poisson processes when they are considered in a steady state of the system. In addition, in NB-IoT, the time intervals of packet arrival are usually on the order of tens of seconds, minutes and hours.

If a traffic model is subject to a Poisson process, the probability that k packets arrive in the duration of t is

$$P_{\text{pos}}\{N = k\} = \frac{(\lambda t)^k}{k!} e^{-\lambda t}, \quad k = 0, 1, 2, \cdots . \tag{1}$$

where the expected time interval of a packet arrival is $\frac{1}{\lambda}$. If $t \ll \frac{1}{\lambda}$, we can assume that there is at most one packet arrive in this duration, and the corresponding probability is λt.

In a NB-IoT network, let the period of one RA attempt be δ, and the average time internal of a packet arrival with UE type i be $\frac{1}{\lambda_i}$. Considering $t = \delta \ll \frac{1}{\lambda_i}$ is satisfied with UE type i, the probability of one packet arrival is $A_i \approx \lambda_i \delta$, and the probability of no packet arrival is $1 - A_i$.

2.2 Contention-Based Random Access Procedure

The data processing of packet buffer follows the principle of "first in first out (FIFO)". When the buffer queue is empty, the UE keeps in an idle state. When there are packets in buffer queue, the UE will initiate a RA procedure if it is not busy. No matter the result is successful or not, the buffer of current packet will be released when the current RA procedure is at an end. If a new packet arrives when the buffer queue is full, this packet will be dropped out. Assuming that the buffer queue with UE type i can contain at most L_i packets ignoring the influence of data size.

Before performing a RA attempt, UEs conduct ACB judgment based on the ACB factor, which is equal to the probability of passing the judgment. UEs that pass the judgment will continue with the RA attempt. UEs that fail in the judgment will wait for some time, and then perform a new ACB judgment. The waiting time is subject to a discrete uniform distribution within the ACB window. With UE type i, let the ACB factor be $P_{\mathrm{ACB}|i}$, and the ACB window be $[W_{\mathrm{ACBmin}|i}, W_{\mathrm{ACBmax}|i}]$.

Assume that the resource allocated to the network consists of M available subcarriers, which is equal to the number of available RA preambles. When performing a RA attempt, the UE selects a RA preamble randomly and transmits it. The preamble may suffer from an extreme channel fading and fail to be received by the base station. If the preamble is received by the base station successfully, the UE will participate in the access contention. In the access contention, if there is no other UE that has transmitted the same preamble, the UE accesses to the network successfully. Otherwise, a collision occurs, and these collided UEs fail in this RA attempt. With UE type i, let the probability of a successful preamble transmission be $P_{\mathrm{d}|i}$, and the probability of a successful access in the contention be $P_{\mathrm{s}|i}$.

If the UE fails in a preamble transmission or an access contention, it will wait for some time, and then perform a new RA attempt. The waiting time is subject to a discrete uniform distribution within the back-off window. The accumulation of RA attempts in one RA procedure is called as RA retransmission, which is shared by both ACB scheme and the back-off scheme. If the number of failed RA retransmission reaches a specific maximum value, the current RA procedure will be considered as failed, and the current packet will be dropped out. With UE type i, let the back-off window be $[0, W_{\mathrm{BO}|i} - 1]$, and the maximum number of RA retransmission be R_i.

3 Analytical Model Based on Markov Chains

In the Markov chain, a moment is defined as a period of RA attempt, and the nodes are in term of a four-dimensional tuple (i, j, r, k) representing the possible states of UE in a RA procedure. Let $i \in [1, N_{\mathrm{class}}]$ denote the index of UE type, $j \in [0, L_i]$ denote the number of packets in buffer queue, $r \in [0, R_i]$ denote the number of RA retransmission, and $k \in [0, W_{\max|i} - 1]$ denote the waiting timer where $W_{\max|i} = \max\{W_{\mathrm{ACBmax}|i}, W_{\mathrm{BO}|i}\}$.

According to the system model, the subchain of one UE type (when i is fixed) has the property of connectivity, which means that the Markov chain will reach a global steady state after a sufficient amount of iterations. Let the state probability in the Markov chain be $\pi(i, j, r, k)$, which is subject to $\sum_{j,r,k} \pi(i, j, r, k) = 1$. When the Markov chain is in steady state, $\pi(i, j, r, k)$ converges to a specific value.

When a packet arrives, it will not be processed until the next period of RA attempt. In other words, the new arrived packet will not affect the process in current RA period. Therefore, the state transition during one RA period can be divided into two independent parts. One is the effect of process in current RA period, the other one is the effect of packet arrival. The probability of state transition can be given by

$$\begin{aligned} &P\{(i, j_1, r_1, k_1) \rightarrow (i, j_2, r_2, k_2)\} \\ &\quad = P_1\{(i, j_1, r_1, k_1) \rightarrow (i, j', r', k')\} \times P_2\{(i, j', r', k') \rightarrow (i, j_2, r_2, k_2)\}. \end{aligned} \tag{2}$$

3.1 State Transition of Process in Current RA Period

For the state transition probability of process in current RA period $P_1\{(i,j_1,r_1,k_1) \to (i,j',r',k')\}$, we discuss the possible situations as follows.

- When the buffer queue is empty, the UE will keep in the idle state.

$$P_1\{(i,0,0,0) \to (i,0,0,0)\} = 1. \tag{3}$$

- The UE succeeds in the RA attempt when the number of RA retransmission doesn't reach its maximum.

$$P_1\{(i,j,r,0) \to (i,j-1,0,0)\} = P_{\mathrm{ACB}|i} \cdot P_{d|i} \cdot P_{s|i}. \tag{4}$$

where $j > 0,\ 0 < r < R_i$.

- The UE fails in the ACB judgment, the preamble transmission or the access contention, and then it should wait some time for the next RA attempt. At this time, the number of RA retransmission doesn't reach its maximum.

$$\begin{aligned} &P_1\{(i,j,r,0) \to (i,j,r+1,k)\} \\ &= u_{\mathrm{ACB}}(k) \cdot (1-P_{\mathrm{ACB}|i}) \cdot \frac{1}{W_{\mathrm{ACBmax}|i} - W_{\mathrm{ACBmin}|i} + 1} \\ &\quad + u_{\mathrm{BO}}(k) \cdot P_{\mathrm{ACB}|i} \cdot (1 - P_{d|i} \cdot P_{s|i}) \cdot \frac{1}{W_{\mathrm{BO}|i}}. \end{aligned} \tag{5}$$

where $j > 0,\ 0 < r < R_i,\ 0 \le k \le W_{\max|i} - 1$, $W_{\max|i} = \max\{W_{\mathrm{ACBmax}|i}, W_{\mathrm{BO}|i}\}$ and

$$u_{\mathrm{ACB}}(k) = \begin{cases} 1,\ k \in [W_{\mathrm{ACBmin}|i} - 1, W_{\mathrm{ACBmax}|i} - 1]; \\ 0,\ \text{otherwise}. \end{cases}$$

$$u_{\mathrm{BO}}(k) = \begin{cases} 1,\ k \in [0, W_{\mathrm{BO}|i} - 1]; \\ 0,\ \text{otherwise}. \end{cases}$$

- The UE performs a RA attempt when the number of RA retransmission reaches its maximum. No matter what the result is, the current RA procedure comes to an end.

$$P_1\{(i,j,R_i,0) \to (i,j-1,0,0)\} = 1. \tag{6}$$

where $j > 0$.

- The UE is waiting for the next RA attempt.

$$P_1\{(i,j,r,k) \to (i,j,r,k-1)\} = 1. \tag{7}$$

where $j > 1,\ r > 1,\ k > 0$.

3.2 State Transition of Packet Arrival

For the state transition of packet arrival $P_2\{(i, j', r', k') \rightarrow (i, j_2, r_2, k_2)\}$, we discuss the possible situations as follows.

- When the UE is in the idle state and the buffer queue is empty, if a new packet arrives, the UE will initiate a new RA procedure. Otherwise, it will be kept in the idle state.

$$P_2\{(i, 0, 0, 0) \rightarrow (i, 0, 0, 0)\} = 1 - A_i. \tag{8}$$
$$P_2\{(i, 0, 0, 0) \rightarrow (i, 1, 1, 0)\} = A_i. \tag{9}$$

- If there are packets in the buffer queue when the current RA procedure is at an end, the UE will initiate a new RA procedure no matter a new packet has arrived or not.

$$P_2\{(i, j, 0, 0) \rightarrow (i, j, 1, 0)\} = 1 - A_i. \tag{10}$$
$$P_2\{(i, j, 0, 0) \rightarrow (i, j + 1, 1, 0)\} = A_i. \tag{11}$$

  where $0 < j < L_i$.
- If the UE is in the process of RA when the buffer queue is not full, the new arrived packet will be stored in buffer.

$$P_2\{(i, j, r, k) \rightarrow (i, j, r, k)\} = 1 - A_i. \tag{12}$$
$$P_2\{(i, j, r, k) \rightarrow (i, j + 1, r, k)\} = A_i. \tag{13}$$

  where $0 < j < L_i$, $r > 1$.
- If the UE is in the process of RA when the buffer queue is full, the new arrived packet will be dropped out.

$$P_2\{(i, L_i, r, k) \rightarrow (i, L_i, r, k)\} = 1. \tag{14}$$

  where $r > 1$.

3.3 Success Rate of Access Contention

Firstly, we consider a general case. Let C denote the probability that a UE participates in an access contention, and we assume that there are totally N UEs deployed. The probability that n UEs is participating in the access contention can be expressed as

$$P_{\text{access}}(n|N, C) = \binom{N}{n} C^n (1 - C)^{N-n}. \tag{15}$$

From the perspective of the base station, when there are M preambles available, the probability that a specific preamble is not transmitted by any UE during a RA period can be expressed as

$$\begin{aligned} P_{\text{exclude}}(N,C) &= \sum_{n=1}^{N} P_{\text{access}}(n|N,C)\cdot(1-\frac{1}{M})^{n} \\ &= \sum_{n=1}^{N}\binom{N}{n}\cdot(C-\frac{C}{M})^{n}\cdot(1-C)^{N-n} \\ &= (1-\frac{C}{M})^{N}. \end{aligned} \tag{16}$$

where we have simplified the expression by using the binomial theorem.

According to the system model, the probability that a UE with UE type i participates in the access contention can be expressed as

$$C_i = \sum_{j=1}^{L_i}\sum_{r=1}^{R_i}\pi(i,j,r,0)\cdot P_{\text{ACB}|i}\cdot P_{d|i}. \tag{17}$$

For a UE with UE type i participating in the access contention, if its preamble has not been transmitted by any other UE, this UE succeeds in the access contention. Therefore, the success rate of the access contention for a UE with UE type i can be expressed as

$$\begin{aligned} P_{s|i} &= \binom{M}{1}\frac{1}{M}\cdot P_{\text{exclude}}(N_i-1,C_i)\cdot\prod_{i'\neq i}P_{\text{exclude}}(N_{i'},C_{i'}) \\ &= (1-\frac{C_i}{M})^{N_i-1}\cdot\prod_{i'\neq i}(1-\frac{C_{i'}}{M})^{N_{i'}} \\ &= (1-\frac{C_i}{M})^{-1}\cdot\prod_{i'}(1-\frac{C_{i'}}{M})^{N_{i'}}. \end{aligned} \tag{18}$$

3.4 Performance Metrics

When the UE succeeds in the RA attempt, the current packet in buffer will be transmitted and processed successfully. The average throughput of UE type i in a RA period is

$$\tau_i = N_i\cdot\sum_{j=1}^{L_i}\sum_{r=1}^{R_i}\pi(i,j,r,0)\cdot P_{\text{ACB}|i}\cdot P_{d|i}\cdot P_{s|i}. \tag{19}$$

The sources of packet dropout consist of three parts. The first one is that the UE fails in the RA procedure, and the current packet is dropped out. The second one is that when the buffer queue is full and the UE is waiting for the next RA attempt, a new packet arrives and it will be dropped out. The third one is that when the buffer queue is full and the UE fails in a RA attempt before reaching the maximum number of RA retransmission, a new packet arrives and it will be dropped out. The average number of dropout packets of UE type i in a RA period is

$$\begin{aligned}\eta_i = N_i \cdot \Bigg(& \sum_{j=1}^{L_i} \pi(i,j,R_i,0) \cdot (1 - P_{\mathrm{ACB}|i} \cdot P_{d|i} \cdot P_{s|i}) \\ & + \sum_{r=2}^{R_i} \sum_{k=1}^{W_{\max|i}-1} \pi(i,L_i,r,k) \cdot A_i \\ & + \sum_{r=1}^{R_i-1} \pi(i,L_i,r,0) \cdot (1 - P_{\mathrm{ACB}|i} \cdot P_{d|i} \cdot P_{s|i}) \cdot A_i \Bigg). \end{aligned} \tag{20}$$

For a packet with UE type i that has been transmitted successfully, the average time delay can be expressed as

$$t_i = \frac{\sum_{j=1}^{L_i} \sum_{r=1}^{R_i} \pi(i,j,r,0) \cdot \left(1 + (r-1)\, \bar{t}_{\mathrm{wait}|i}\right)}{\sum_{j=1}^{L_i} \sum_{r=1}^{R_i} \pi(i,j,r,0)}, \tag{21}$$

where $\bar{t}_{\mathrm{wait}|i}$ is the average waiting time for a RA retransmission, and it can be expressed as

$$\begin{aligned}\bar{t}_{\mathrm{wait}|i} = & \frac{1}{1 - P_{\mathrm{ACB}|i} \cdot P_{d|i} \cdot P_{s|i}} \\ & \times \left((1 - P_{\mathrm{ACB}|i}) \cdot \frac{W_{\mathrm{ACBmax}|i} + W_{\mathrm{ACBmin}|i}}{2} \right. \\ & \left. + P_{\mathrm{ACB}|i} \cdot (1 - P_{d|i} \cdot P_{s|i}) \cdot \frac{W_{\mathrm{BO}|i} + 1}{2} \right). \end{aligned} \tag{22}$$

The packet passing rate of UE type i can be expressed as

$$\sigma_i = \frac{\tau_i}{\tau_i + \eta_i}, \tag{23}$$

which denotes the proportion of packets processed successfully.

4 Optimization of System

In a heterogeneous network, we not only need to improve the overall performance of the system, but also need to avoid severe performance degradation of any UE types. In other words, we should consider the fairness among UE types. In this paper, the Max-Min fairness (MMF) is adopted to indicate the fairness of packet passing rate among UE types, which can be expressed as

$$F = \frac{\min\{\sigma_i\}}{\max\{\sigma_i\}}. \tag{24}$$

where $F \in [0, 1]$. When F is close to 1, the system will be considered as fair.

To improve the performance of system comprehensively when taking account of the fairness, we configure parameters for each UE type according to its characteristics. Hence, we adopt an optimization criterion, where the overall throughput $\sum_i \tau_i$ is maximized on condition that the fairness metric F is above a prerequisite threshold.

When the number of UE types increases, the searching space of parameter configuration will grow exponentially. Therefore, it is impractical to perform an exhaustive search to obtain the optimal configuration. In this paper, we propose an optimization algorithm based on simulated annealing, which can obtain an approximate optimal solution with an acceptable complexity.

Let x_n denote a solution of parameter configuration in the n-th iteration within the searching space X, and $E_m(x_n)$ denote the metrics to be maximized by means of simulated annealing. $E_1(x_n)$ is the condition function with a prerequisite threshold θ, which denotes the MMF indicator F. $E_2(x_n)$ is the objective function, which denotes the overall throughput $\sum_i \tau_i$. When performing simulated annealing, let N_{SA} be the total number of iterations, α be the fading factor of temperature, and $T_1(0), T_2(0)$ be the initial temperatures. The procedure of the optimization is shown in Algorithm 1.

Algorithm 1. Optimization based on simulated annealing

Require: X, α, $T_1(0)$, $T_2(0)$, N_{SA}.

generate an initial solution $x_0 \in X$ randomly.
for $n = 0$ to $N_{\mathrm{SA}} - 1$ **do**
  generate a new solution $x_n^* \in X$ through a random perturbation of x_n.
  if $E_1(x_n) < \theta$ or $E_1(x_n^*) < \theta$ **then**
    $m = 1$
  else
    $m = 2$
  end if
  if $E_m(x_n^*) > E_m(x_n)$ **then**
    $x_{n+1} = x_n^*$
  else
    $T_m(n) = \alpha^n \cdot T_m(0)$
    generate $p \in [0, 1]$ randomly
    if $p < \exp(-\frac{|E_m(x_n^*) - E_m(x_n)|}{T_m(n)})$ **then**
      $x_{n+1} = x_n^*$
    else
      $x_{n+1} = x_n$
    end if
  end if
end for
return $x_{N_{\mathrm{SA}}}$.

5 Simulation and Discussion

To simulate the RA procedure in NB-IoT, a system-level simulator based on slotted ALOHA protocol is built in MATLAB platform, and then the performance metrics of the system can be obtained by means of Monte Carlo method.

For ease of analysis, we consider the situation that there is only a cellular cell deployed in the network, where the common configuration of resource allocation is $M = 36$, $\delta = 1280$ ms. In the heterogeneous network, there are $N_{\text{class}} = 3$ UE types deployed at a ratio of $N_1 : N_2 : N_3 = 3 : 4 : 5$. The characteristics of each UE type is shown in Table 1

Table 1. The characteristics of UE type i

Characteristic	Type 1	Type 2	Type 3
λ_i^{-1}	60 s	90 s	120 s
$P_{\text{d}\|i}$	70%	95%	95%
L_i	3	3	3

5.1 Verification of Analytical Model

To prove the authenticity of the proposed analytical model in a heterogeneous network, we compare the performance metrics between the analytical model and the simulator in the same configuration. To emphasize the differences among UE types, we provide a static configuration for each UE type individually, which is shown in Table 2.

Table 2. Static configuration for UE type i

Parameter	Type 1	Type 2	Type 3
R_i	5	7	10
$W_{\text{BO}\|i}$	30	20	20
$P_{\text{ACB}\|i}$	95%	65%	50%
$W_{\text{ACBmin}\|i}/W_{\text{ACBmax}\|i}$	24/32	18/24	18/24

Figure 1 shows the throughput of each UE type. The throughput grows linearly when $N < 1000$, and then turns to decrease when $N > 1500$. Figure 2 shows the packet dropout of each UE type. The packet dropout increases slowly when $N < 1000$ and then grows linearly when $N > 1500$. It is because when the network is uncrowded ($N < 1000$), UEs can almost access to the network and the new arrived packets can be processed in time instead of being dropped out.

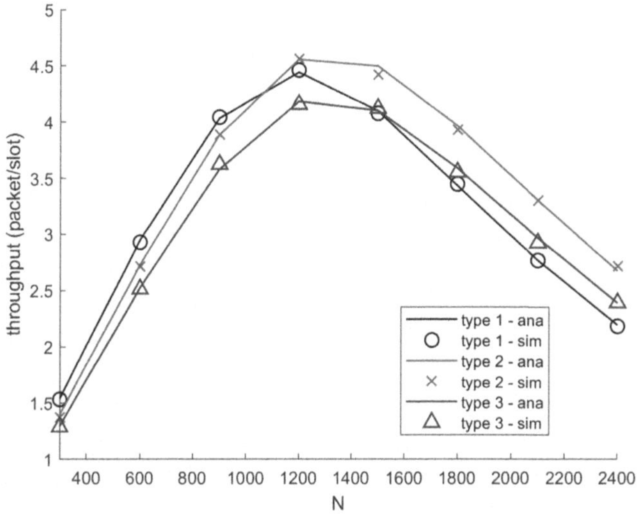

Fig. 1. Throughput of each UE type

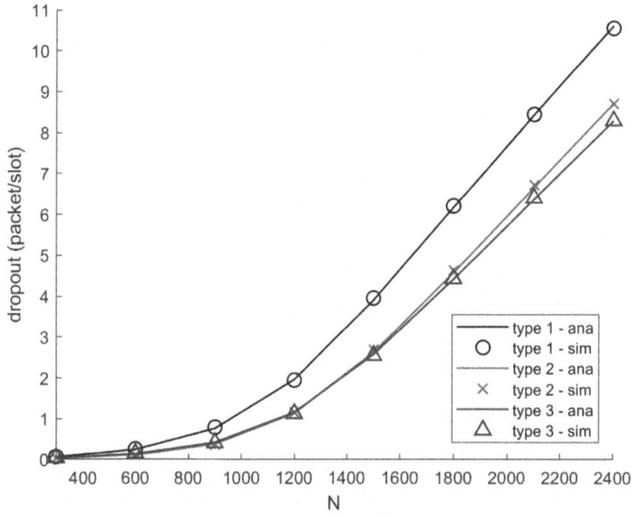

Fig. 2. Dropout of each UE type

When the network is crowded ($N > 1500$), an intenser contention will result in a larger proportion of UEs that fail in the RA procedure.

Figure 3 shows the average delay of each UE type. The average delay keeps increasing when UEs need to conduct more RA attempts in ACB scheme and the back-off scheme, and it will converge to a static value due to the limitation of R_i.

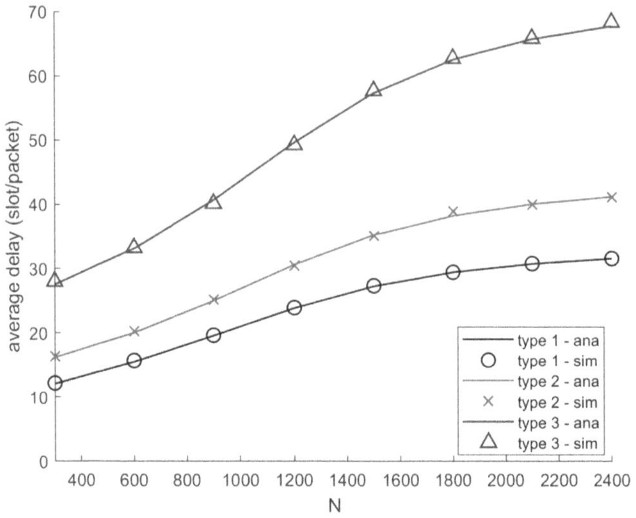

Fig. 3. Average delay of each UE type

The performance metrics of each UE type have the same tendency as N increases, because every UE is on an equality when participating in the contention of RA. The performance metrics of each UE type derived from the proposed analytical model are nearly equal to that obtained from the simulator. The relative error between them is within 2%. This result shows that the proposed analytical model matches the real system exactly, so it is reliable to optimize the RA procedure based on the analysis of analytical model. Compared to the high complexity of Monte Carlo method in the simulator, the analytical model achieves a low complexity, where the model can converge after only a few iterations.

5.2 Optimization of the System

To evaluate the effect of optimization, two strategies based on simulated annealing are examined. One is the proposed strategy which makes a trade off between the overall throughput and the fairness, and the other one is a general strategy that only maximizes the overall throughput. To initialize the simulated annealing algorithm, let $\theta = 0.85$, $T_1(0) = 0.3$, $T_2(0) = 2$, $\alpha = 0.98$, $N_{\mathrm{SA}} = 200$. The searching space of parameter configuration is shown in Table 3.

The result of the overall throughput and the fairness indicator after optimization are shown in Fig. 4 and 5 respectively. When the network is uncrowded ($N < 1000$), the overall throughput of the two strategies increases at the same rate, and the fairness indicators of them are both close to 1. When the network is crowded ($N > 1500$), in the general strategy, the overall throughput is kept in a high level, while the fairness indicator is within 50%–60%. In the proposed strategy, the overall throughput decreases slightly and still be in a relative high

Table 3. Searching space of configuration

Parameter	Searching space
R_i	$4, 5, 6, \cdots, 12$
$W_{\text{BO}\|i}$	$5, 6, 7, \cdots, 32$
$P_{\text{ACB}\|i}$	$50\%, 55\%, 60\%, \cdots, 100\%$
$W_{\text{ACBmax}\|i}$	$15, 16, 17, \cdots, 36$
$W_{\text{ACBmin}\|i}$	$\lfloor 0.75 \cdot W_{\text{ACBmax}\|i} \rfloor$

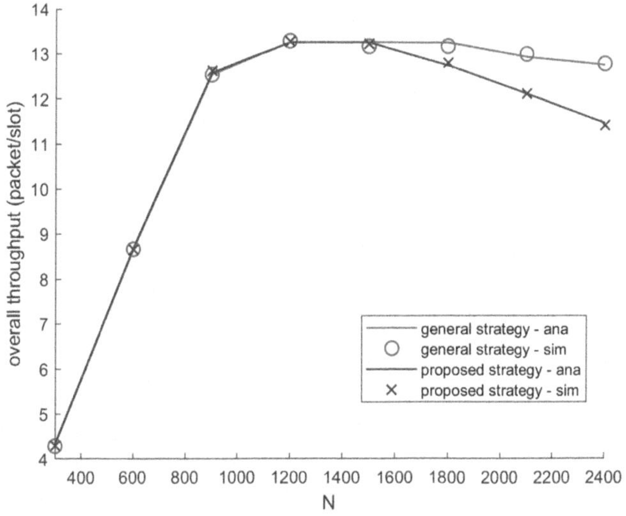

Fig. 4. Overall throughput after optimization

level, while the fairness is guaranteed ($F > 85\%$). It means that the proposed strategy does well in making a trade off between the throughput and the fairness.

For further analysis, the packet passing rates of each UE type in the two strategies are shown in Fig. 6 and 7 respectively. When $N < 1000$, the packet passing rate of each UE type can be close to 1, so the problem of fairness is ignorable. When $N > 1000$, differences between the two strategies occur. In the general strategy, the packet passing rate of UE type 1, where packets arrive more frequently, decreases faster than that of other UE types. It leads to a serious problem of fairness. It is because to maximize the overall throughput, the system prefers to restricting the access of UEs with a higher amount of traffic, which can ease congestion more effectively at the price of the performance reduction of these UEs. In the proposed strategy, the packet passing rates of UE types are close to each other. It shows that to guarantee the fairness, the UE types with good performance need to make concession to other UE types on resource allocation, which results in a slight decline of the overall throughput.

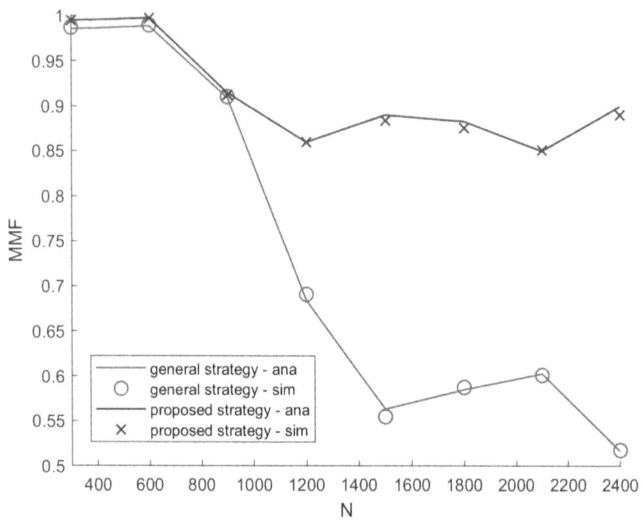

Fig. 5. Fairness indicator after optimization

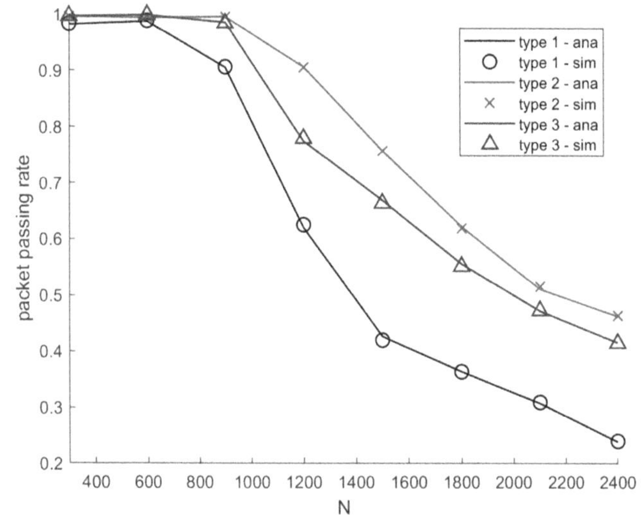

Fig. 6. Packet passing rate in the general strategy

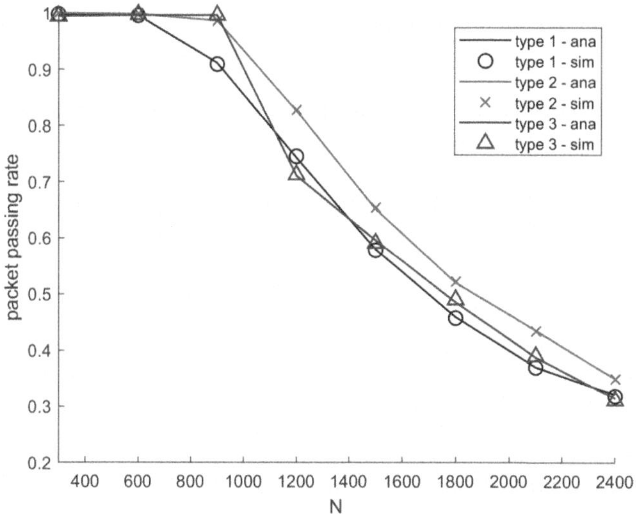

Fig. 7. Packet passing rate in the proposed strategy

6 Conclusion

In this paper, considering a case of heterogeneous networks, a model based on Markov chains is proposed to evaluate the performance of contention-based RA procedure in NB-IoT. Simulation results show that the proposed analytical model is closely approximate to the performance metrics of real systems. Based on the analysis of the model, an optimization algorithm based on simulated annealing is proposed, which configures parameters for each UE type individually and makes a trade off between the overall throughput and the fairness. Simulation results show that the proposed algorithm can find an approximate optimal system configuration with an acceptable complexity, which keeps the overall throughput at a relatively high level under the prerequisite of guaranteeing the fairness.

References

1. Rastogi, E., Saxena, N., Roy, A., Shin, D.R.: Narrowband internet of things: a comprehensive study. Comput. Netw. **173**, 107209 (2020)
2. Pennacchioni, M., Di Benedette, M.-G., Pecorella, T., Carlini, C., Obino, P.: NB-IoT system deployment for smart metering: evaluation of coverage and capacity performances. In: AEIT International Annual Conference, Cagliari, Italy, pp. 1–6 (2017)
3. "Technical Specification Group Radio Access Network; Evolved Universal Terrestrial Radio Access (E-UTRA); Medium Access Control (MAC) protocol specification (Release 14)", 3GPP, TS 36.321 V14.6.0 (2018)
4. Kanj, M., Savaux, V., Le Guen, M.: A tutorial on NB-IoT physical layer design. IEEE Commun. Surv. Tutor. **22**(4), 2408–2446 (2020)

5. "Technical Specification Group Radio Access Network; Study on RAN Improvements for Machine-type Communications; (Release 11)", 3GPP, TR 37.868 V11.0.0 (2011)
6. Ramineni, C.: Impact of access class barring in narrow band internet of things. IET Wirel. Sens. Syst. **9**(6), 399–404 (2019)
7. Zhao, Y., Liu, K., Yan, H., Huang, L.: A classification back-off method for capacity optimization in NB-IoT random access. In: 2017 11th IEEE International Conference on Anti-counterfeiting, Security, and Identification (ASID), Xiamen, China, pp. 104–108 (2017)
8. Harwahyu, R., Cheng, R.-G., Wei, C.-H., Sari, R.F.: Optimization of random access channel in NB-IoT. IEEE Internet Things J. **5**(1), 391–402 (2018)
9. Harwahyu, R., Cheng, R.-G., Liu, D.-H., Sari, R.F.: Fair configuration scheme for random access in NB-IoT with multiple coverage enhancement levels. IEEE Trans. Mob. Comput. **20**(4), 1408–1419 (2021)
10. Harwahyu, R., Cheng, R.-G., Tsai, W.-J., Hwang, J.-K., Bianchi, G.: Repetitions versus retransmissions: tradeoff in configuring NB-IoT random access channels. IEEE Internet Things J. **6**(2), 3796–3805 (2019)
11. Jiang, N., Deng, Y., Kang, X., Nallanathan, A.: Random access analysis for massive IoT networks under a new spatio-temporal model: a stochastic geometry approach. IEEE Trans. Commun. **66**(11), 5788–5803 (2018)
12. Sun, Y., Tong, F., Zhang, Z., He, S.: Throughput modeling and analysis of random access in narrowband Internet of Things. IEEE Internet Things J. **5**(3), 1485–1493 (2018)
13. Ramineni, C., Venkatesh, T.G., Bommisetty, L.: Performance evaluation of random access in narrow band Internet of Things. Comput. Netw. **218**, 109399 (2022)
14. Xia, T., Wang, M.M., Jiang, C., Zhang, J., Wang, L., You, X.: Practical machine-type communication for energy Internet of Things: an introduction. IEEE Commun. Stand. Mag. **3**(1), 48–59 (2019)
15. Navarro-Ortiz, J., Romero-Diaz, P., Sendra, S., Ameigeiras, P., Ramos-Munoz, J.J., Lopez-Soler, J.M.: A survey on 5G usage scenarios and traffic models. IEEE Commun. Surv. Tutor. **22**(2), 905–929 (2020)

Research on Routing Algorithm of Space-Air Integrated Networks

Kexin Zhang[1], Zhongyu Yin[1], Tong Liu[2], and Shuo Shi[1(✉)]

[1] Harbin Institute of Technology, Harbin 150001, Heilongjiang, China
crcss@hit.edu.cn

[2] Norinco Group Air Ammunition Research Institute, Harbin, China

Abstract. Satellite communication networks have many outstanding advantages over traditional communication systems, such as terrestrial cellular networks, and will play a central role in next-generation wireless communication systems. Low-orbit satellite communication systems have great potential and value due to their advantages such as lower transmission delay and higher communication quality. However, due to the dynamic nature of satellite constellation topology, load imbalance and node/link failure problems, the design of routing algorithms puts forward high requirements. This paper analyses the inclined orbit topology characteristics of low earth orbit (LEO) satellite network and the influence of phase factor according to the characteristics of LEO satellite network, and on this basis, analyses the advantages and disadvantages of the existing routing protocols in conjunction with the challenge of the design of routing algorithms for LEO satellite network.

Keywords: Satellite communication · LEO · Routing algorithms

1 Sources and Significance of the Research

With the advent of the post-5G era, customised, high-bandwidth, low-latency services are now possible over terrestrial networks. However, geographic constraints have resulted in terrestrial networks covering only about 20 per cent of land and 6 per cent of the Earth's surface area, making it impossible for terrestrial networks alone to meet the expectations of ubiquitous communications in the 6G era. Against this backdrop, satellite communications have emerged as a reliable means of communication. Compared with traditional communication systems such as terrestrial cellular networks, satellite communication systems have the following outstanding advantages: (i) wide coverage and high reliability, (ii) satellite communication systems have huge processing power, (iii) communication costs are insensitive to distance. Based on the advantages of satellite communication systems mentioned above, the global network information industry generally believes that satellite communication networks have great potential for application development and are expected to play a central role in the next-generation wireless communication system.

In satellite communication networks, satellites can be classified into low orbit, medium orbit (MEO) and geostationary orbit (GEO) satellites according to their orbital

X. Li et al. (Eds.): 6GN 2024, LNICST 615, pp. 45–54, 2025.
https://doi.org/10.1007/978-3-031-93193-2_4

altitude. Compared with the other two satellite communication systems, LEO satellites have significant advantages due to their lower orbital altitude, and LEO satellite communication systems have advantages such as lower transmission delay and higher communication quality. In addition, the LEO constellation network has the advantages of dynamic coverage, high scalability, and strong anti-interference ability, and it can be closely integrated with the ground communication infrastructure to jointly construct a seamless global coverage communication platform with air, ground, and satellite synergy, and provide key technical support in many fields, which has great potential and value [1].

Routing algorithm design is an important issue in the networking technology of satellite networks, which directly affects the service quality of user services as well as the resource utilisation of the system. Since the orbital operation period of LEO satellites is shorter and the network topology changes faster, it is necessary to ensure the communication quality and adapt to the dynamically changing network topology during the communication process, in addition, the load balancing and anti-destruction performance also need to be considered in the algorithm. This makes the routing problem a hot issue in the research of low-orbit satellite networks.

2 Development of LEO Satellite Network

Since 2015, a large number of low-orbit megaconstellation programmes have been proposed, including representative programmes such as OneWeb, Starlink, and Telesat.

(1) OneWeb

OneWeb is a broadband LEO constellation programme proposed by the UK-based OneWeb, which plans to launch 716 satellites at an orbital altitude of 1,200 km to form a hybrid polar/inclined orbit constellation to provide broadband Internet access. By the end of 2021, OneWeb has deployed nearly 400 satellites and plans to expand the constellation to 6,372 satellites in the future to increase the coverage density in low and mid-latitudes. OneWeb will be involved in a number of application areas [2], including: 1) maritime, aviation, and land mobility; 2) enterprise and consumer broadband access; 3) emergency communications; and 4) cellular backhaul.

(2) Starlink

Starlink [3] was proposed by SpaceX, which has filed several applications and amendments with the federal communications commission (FCC), and its constellation consists of several phases. The initial plan is to launch 4,408 satellites operating at an altitude of 540–570 km to provide Internet services to users around the world, with the primary market being broadband Internet access for individual consumers, which is expected to provide transmission rates of up to 1Gbps per user in the future. Benefiting from leading technologies such as bulk satellite manufacturing, rocket reuse, and multi-star launches, Starlink has become a leader in the emerging low-orbit constellations, with over 1,900 satellites deployed by the end of 2021, completing the deployment of the constellation in the 550 km orbital layer. Next, Starlink will deploy additional orbital layer constellations and begin to equip them with laser interplanetary links. In the future, Starlink also plans to increase the total constellation size to 42,000 satellites.

(3) Telesat

Telesat is a broadband Internet constellation proposed by Telesat, a Canadian satellite operator, which consists of two sets of orbits, one polar, with an inclination of 98.98° and an altitude of 1015 km, and the other inclined, with an inclination of 50.88° and an altitude of 1,325 km, to combine global coverage with hotspot coverage using a combination of polar and inclined orbits. Telesat aims to provide services to commercial, government and military users, including mobile backhaul, aeronautical and maritime broadband, etc. Telesat is not involved in the consumer broadband market. Currently, Telesat has launched only one test satellite.

3 LEO Satellite Network Basic Concepts

(1) Satellite Orbit

The Keplerian orbit element is commonly used to describe a satellite orbit and the position of the satellite in the orbit and contains the following parameters:

Orbital semi-long axis: Half of the long axis of a satellite's elliptical orbit.

Orbital eccentricity: a measure of the degree of flattening of an ellipse, defined as the ratio of the distance between the two foci of the ellipse to the length of the long axis.

Orbital inclination: the inclination of the planet's orbital plane to the equatorial plane; the angle measured counterclockwise from the equatorial plane to the planet's orbital plane at the ascending node.

Right ascension of the ascending node (RAAN): the satellite orbit and the equatorial plane intersect at two points (except for orbital inclination 0°), respectively known as the ascending node and descending node. The satellite moves from south to north at the ascending node and from north to south at the descending node. Ascending node declination is the angle from the equinox to the ascending node with the Earth's centre as the vertex, and takes the value of 0°–360°.

Perigee angle: Perigee is the point in the elliptical orbit of a spacecraft around the Earth that is closest to the centre of the Earth. The perigee angle is the angle between the perigee and the ascending node of the orbit with respect to the centre of the Earth. It is measured from the ascending node to the perigee in the direction of satellite motion, and its value ranges from 0° to 360°.

True Perigee Angle: The true perigee angle determines the exact position of the satellite in orbit at the time of the calendar element and takes the value of 0°–360°, where perigee corresponds to 0° and apogee to 180°.

Circular orbits are commonly used in existing LEO satellite systems because they provide consistent coverage of the ground. For circular orbits, the perigee amplitude is 0, the eccentricity is 0, and the half-length axis is equal to the radius of the orbit.

Other orbital elements are shown in Fig. 1.

(2) Orbit Type

According to the satellite orbit altitude, satellites can be divided into LEO satellites, MEO satellites and GEO satellites, and different orbit altitudes have different characteristics.

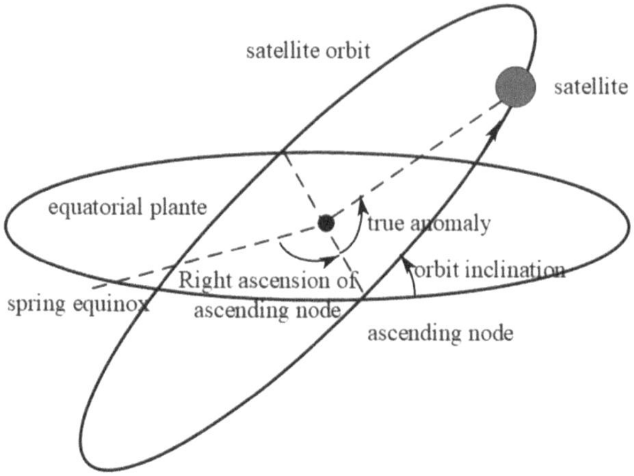

Fig. 1. Satellite orbital elements.

LEO satellites are located at an altitude of 400–2000 km. Due to the lower orbital altitude of LEO satellites, the transmission delay is relatively shorter and the path loss is also smaller. This enables LEO satellites to provide more efficient and faster communication services. By deploying multiple LEO satellites, a global coverage network can be formed, but due to the fast speed of LEO satellites orbiting around the orbit, the Doppler shift of the star-ground link is large, and the ground terminals need to switch satellites frequently.

MEO satellites have an orbital altitude of 2,000–20,000 km, an operation period of about 12 h and a sub-satellite track that can cover the whole world, which is suitable for national measurement and control. Since the orbit altitude of MEO satellites is moderate, their launching cost is relatively low and they have strong coverage capability in high-latitude areas. The switching frequency between ground terminals and satellites is reduced but the link delay is increased accordingly.

GEO satellites are located in the equatorial plane 35786 km away from the earth, showing a 'static' state relative to the earth's surface, and can realise global communication except for the poles. Most of the world can be covered by three satellites, but due to the long distance, a large aperture antenna is required and the signal transmission delay is large, which may affect the quality of communication.

(3) Walker's Constellation

The Walker constellation is a constellation of satellites distributed in a number of circular orbits, each with the same orbital altitude and inclination, with orbits evenly spaced and with satellites evenly spaced in each orbit.

Depending on the angle of inclination of the orbit, Walker constellations can be classified into polar orbit constellations and inclined orbit constellations. Walker constellations with orbital inclination close to 90° are called polar constellations or Walker star constellations, as shown in Fig. 2a), and the orbits are uniformly spaced along 180° of the equator, so they are also called π constellations [4]. The advantage of polar constellations is that they provide seamless global coverage. However, due to the existence of the reverse slit, satellites on both sides of the reverse slit cannot establish interstellar links, which is prone to waste of resources. Walker constellations with orbital inclination less than 90° and more than 0° (usually 40° to 80°) are called inclined orbit constellations or Walker delta constellations, as shown in Fig. 2b), and the orbits are uniformly distributed along the equator within a 360-degree range, and are therefore also called 2π constellations, whose advantage lies in providing better coverage for the mid- to low-latitude areas where the population is more concentrated. The advantage of the inclined orbit Walker constellation is that it provides better coverage of the more populated low and mid-latitude regions. The inclined orbit Walker constellation has more uniform ground coverage and no reverse slit, and the inter-orbit interstellar link will not be closed, which makes it possible to make full use of the satellite resources provided that the number of satellites is the same. Therefore, the following section focuses on analysing the topological characteristics of the inclined Walker constellation.

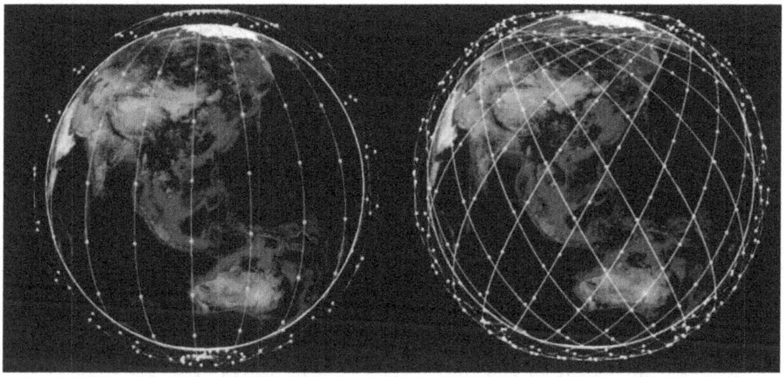

a) Polaris Constellation b) Tilted Orbit Constellation

Fig. 2. Walker constellation.

4 Topological Analysis of Tilt-Orbit Constellations

4.1 Topological Stability

The interstellar links in the LEO satellite constellation can be divided into intra-orbit Inter-Satellite link (ISL) and inter-orbit ISL. Intra-orbit ISL connects satellites in the same orbit in front of and behind each other. Since these satellites operate in the same orbit, the relative positions between them are fixed, so the intra-orbital ISL can maintain a stable connection status. Inter-orbit ISLs, on the other hand, connect satellites in left and right neighbouring orbits. As the satellites move in their respective orbits, the relative positions between them change, which results in the length and pointing of the inter-orbit ISLs also changing. Thus, the dynamics of the network topology is mainly caused by the inter-orbit ISL. In order to accurately describe this variation, azimuth and pitch angles are commonly used to define the pointing of links. The azimuth describes the orientation of the link in the horizontal plane, while the pitch angle describes the angle between the link and the horizontal plane.

The authors in [5] verifie the topological stability of the tilted-orbit constellations, and investigate the angular velocity variation of the pitch and azimuth angles of the inter-orbit ISLs in the Starlink v0, Starlink v2, Iridium, and OneWeb constellations. Starlink v0 denotes the constellations in the first application submitted to the FCC by Starlink, and Starlink Starlink v0 represents the constellation in Starlink's first application to the FCC, and Starlink v2 represents the constellation in Starlink's second revised application to the FCC. It is found that the dynamics of the inter-orbit ISL is mainly caused by the change in azimuth angle, while the change in pitch angle is small. The dynamic range of azimuthal angular velocities is $-0.06°/s$ to $-0.06°/s$ for the inclined constellations Starlink v0 and Starlink v2, and $-0.1°/s$ to $0.1°/s$ for the polar constellations Iridium and OneWeb, showing that the inclined constellations are more dynamic than the polar constellations in terms of azimuthal angular velocities. It can be seen that the dynamic range of azimuthal angular velocities of the inclined-orbit constellation is smaller than that of the polar-orbit constellation, which helps to reduce the antenna tracking capability and to consider the inter-orbit ISL as not disconnected, so that the topology of the network can be kept stable.

4.2 Influence of the Phase Factor

In the design of the Walker constellation, the precise configuration of the constellation configuration parameters (e.g., orbital plane and phase factor, etc.) plays a decisive role in the overall performance and characteristics of the constellation. In particular, the setting of the phase factor has a significant impact on the following three aspects:

(1) Minimum interstellar distance: The minimum distance between stars is a key safety parameter during the operation of a constellation. Inappropriate configuration of the phase factor may result in too close distances between stars or even overlapping orbits, which significantly increases the risk of collision.
(2) Dynamics of inter-orbit links: The configuration of the phase factor also has a direct impact on the range of dynamic adjustments during the establishment of heterorbit interplanetary links. If the phase factor is not set reasonably, the inter-orbit link may require greater angular adjustment when it is established, which puts forward higher technical requirements for the rotation and alignment capability of the on-board laser payload.
(3) ISL hops in routing algorithms: In the design of routing algorithms for constellations, the setting of phase factors has a direct impact on the number of ISL hops. By optimising the phase factor configuration of the constellation, the average ISL hop count of the whole network as well as the users in a specific area can be effectively reduced, thus improving the communication efficiency and network performance.

The authors in [6] give Starlink v2 in different phase factors, the minimum distance between any star better several groups of phase factors: phase factor 53, can achieve the minimum distance between any star 57.0 km; phase factor 45, can achieve the minimum distance between any star 57.0 km; further, the phase factor 53, 45 two groups of constellations configuration, the construction of the chain of heterorbital interstellar relative movement Further, under the three sets of constellation configurations of phase factor 53 and 45, the relative motion speeds during the establishment of the interstellar chain in different orbits are simulated and analysed as follows: for phase factor 53, the maximum relative motion speed of the interstellar chain in different orbits is 0.31 km/s, and for phase factor 45, the maximum relative motion speed of the interstellar chain in different orbits is 0.20 km/s. A theoretical model was proposed in reference [7] to accurately estimate the hops of interplanetary links between terrestrial users on inclined orbits by analyzing interplanetary link paths. Simulation results show that optimising the constellation phase factor can reduce the average ISL hop count during routing. Since other constellation parameters are determined after comprehensive balancing and cannot be easily changed, optimising F is a promising and practical method to reduce the ISL hop count. The selection of the phase factor needs to be considered in the combination of the above three factors.

5 LEO Satellite Network Routing Algorithm

LEO satellite networks are characterised by topological dynamics, uneven service distribution and resource constraints, which pose a series of challenges to the design of routing algorithms and require the design of customised routing algorithms. Domestic and foreign scholars have studied the routing algorithms in LEO satellite networks from several perspectives [8]. In this section, we will review the existing routing algorithms for LEO satellite networks in terms of topology dynamics, load balancing, and destruction resistance, and point out the problems of existing routing algorithms.

5.1 Routing Algorithms for Dynamic Topology Changes

The most important feature that distinguishes LEO satellite networks from terrestrial networks is the time-varying nature of the topology. On the one hand, the distance of inter-orbit interstellar links changes continuously with the motion of satellites, and on the other hand, in polar-orbit constellations, when satellites move to the polar region, the inter-orbit interstellar links are temporarily disconnected due to the fast relative motion speed, and re-established when the satellites leave the polar region. However, due to the movement of satellites along the predetermined orbits, the change of the topology of the LEO satellite network is again regular and predictable, and thus this characteristic can be fully utilised for the design of routing algorithms. There are two main solution ideas for the topology dynamics of satellite networks: virtual topology strategy and virtual node strategy.

The virtual topology strategy creates multiple static virtual topologies by exploiting the periodicity of satellite motion, dividing the large time period into multiple smaller time segments, and assuming that the network topology remains unchanged during each time segment. A discrete-time dynamic virtual topological routing (DT-DVTR) routing algorithm was proposed in reference [9], which uses the idea of virtual topology to deal with complex dynamic changes in satellite networks. The algorithm uses the shortest path algorithm to calculate the set of paths between any two satellites in each static topology network, and selects the path with the least number of switches in adjacent time slices as the principle, and chooses the path with the least number of switches as the final transmission path. However, the DT-DVTR algorithm is less resistant to destruction, and when a satellite node fails, it has a greater impact on the entire satellite constellation network. Increasing the number of time slices will cause larger on-planet storage overheads on the satellite while improving the performance of the network in transmitting data.

The virtual node strategy logically numbers the positions of the satellite nodes in the entire satellite system to form a virtual network with a fixed topology covering the whole world, in order to shield the changes caused by the satellite movement. During satellite operation, the real satellite establishes a mapping relationship with the nearest virtual node by judging its positional relationship with each virtual node in the network, thus simplifying the routing decision-making process. The datagram routing algorithm (DRA) algorithm [10] is the earliest routing algorithm that adopts the concept of virtual node, and based on the position of the current virtual node and the destination virtual node, it firstly finds out the forwarding direction corresponding to the path with the smallest number of hops, and then the forwarding direction corresponding to the path with the minimum forwarding direction corresponding to the delay path. This algorithm improves the destruction resistance and the on-board storage overhead of the satellite is small.

5.2 Load Balancing Routing Algorithm

The LEO satellite network serves users around the world, and due to the uneven distribution of global services, the traffic carried by satellite links is not balanced, and some links in the network carry too much traffic, while other parts of the links are less loaded, which leads to local network congestion, increases queuing delay and packet loss, and reduces the overall performance of the network. Therefore, the routing algorithm needs to be designed with load balancing in mind to optimise the throughput of the network.

Since the LEO satellite network has a more regular mesh topology, there are multiple paths with similar delay between a pair of satellite nodes, thus providing a basis for the design of load balancing routing algorithms [11]. The finite state automaton (FSA) [12] algorithm is a typical global load balancing routing algorithm, which divides the system operation period into multiple time slices, and the topology within each time slice is treated as fixed invariant, and link allocation is based on the delay in the fixed topology, this algorithm lacks a better performance in relieving link congestion in a timely manner. The explicit load balancing algorithm [13] is a typical local load balancing routing algorithm. It reduces the service sending rate when the link buffer utilisation exceeds the threshold and finds alternate paths to relieve the current satellite node congestion. However, this algorithm shunts in a simple way and has limited performance in relieving link congestion. The authors in [14] proposed a compact explicit multi-path routing algorithm, which firstly proposes a track spokesman mechanism, where a track spokesman is elected within each track to collect link information within the track, and then the track spokesmen pass link information to each other, so that each track spokesman finally gets the link status of the whole network. Then, k shortest paths are calculated for each source/destination node to each other. The data forwarding process uses source routing to carry the routes in the data and distribute the traffic to the k paths for load balancing purpose.

5.3 Destruction-Resistant Routing Algorithm

In LEO satellite networks, the satellite Internet is vulnerable to attacks and interference due to the circular visibility of satellite orbits and the fixed nature of orbits. In addition, the effects of the cosmic space environment and device aging will cause the satellite to fail, and the interstellar link associated with it will be broken, thus triggering changes in the topology of the satellite network. Routing algorithms should have a certain degree of robustness or resistance to destruction, and still be able to ensure the correct delivery of data packets as much as possible when the network topology is abnormal. Improving the network anti-interference ability and robustness is also a hot topic in satellite Internet routing research in recent years.

The DRA algorithm employs a local roundabout type of strategy where packets are transmitted along the suboptimal next hop when the optimal next hop is not available [10]. However, this routing mechanism utilises only local information and tends to result in non-optimal paths, thus introducing additional path delay. The authors in [15] proposed a destruction-resistant dynamic routing algorithm that divides the satellite network into clusters, optimises the switching effect by employing a boundary satellite source routing scheme, and enhances the system survivability in the event of interstellar link failure. Since the anti-destructive dynamic routing algorithm needs to update the link state, it will incur additional transmission overhead, based on this literature [16] further proposed an anti-destructive routing algorithm for LEO satellite networks, in order to automatically avoid ineffective inter-satellite links and reroute them at a minimal cost. The authors in [17] proposed a heterogeneous satellite Internet anti-destructive routing scheme based on game theory and deep learning, modelling the routing anti-destructive problem as a hierarchical anti-destructive Stackelberg game, and the resulting routing strategy has lower overhead and better anti-destructive performance.

In summary, existing routing algorithms for LEO satellite networks have been studied for topology dynamics, load balancing, and destruction resistance, yet there are still some unsolved problems. With the increase in the scale of satellite networks, existing algorithms usually lack scalability and introduce higher signalling overhead when coping with link failures in the network, while some algorithms have some impact on the path delay performance while reducing the overhead.

6 Conclusion

This paper takes the routing problem of LEO satellite network as the target, analyses the topological characteristics of the inclined orbit of LEO satellite network as well as the influence of phase factor selection, and on this basis, analyses the advantages and disadvantages of the existing routing protocols in conjunction with the challenge of the design of routing algorithms for LEO satellite network. Existing literature has not yet designed a targeted routing algorithm for the giant inclined-orbit constellations by fully studying their topological characteristics and the influence of phase factors.

Acknowledgement. This work has received support from the National Natural Science Foundation of China under Grant 62171158, National Key Laboratory of Air-based Information Perception and Fusion and Aeronautical Science Foundation of China under Grant 20230001004002 and the Fundamental Research Funds for the Central Universities under Grant 2242022k60006.

References

1. Su, Y., Liu, Y., Zhou, Y., Yuan, J., Cao, H., Shi, J.: Broadband LEO satellite communications: architectures and key technologies. IEEE Wirel. Commun. **26**(2), 55–61 (2019)
2. Yuxuan, G., Yue, L., Penghui, S.: Research status of typical satellite communication systems. In: 2021 19th International Conference on Optical Communications and Networks (ICOCN), Qufu, pp. 1–3. IEEE (2021)
3. Kuzu, L., Yagli, A.F., Gokten, M., Yanikgonul, V.: Reliability analysis of TUSAT satellite communication payload. In: 2012 IEEE First AESS European Conference on Satellite Telecommunications (ESTEL), Rome, pp. 1–5. IEEE (2012)
4. Ferreira, A., Galtier, J., Penna, P.: Handbook of Wireless Networks and Mobile Computing, 1st edn. Wiley-Blackwell, New York (2002)
5. Gavish, B., Kalvenes, J.: The impact of intersatellite communication links on LEOS performance. Telecommun. Syst. **8**(2), 159–190 (1997)
6. Ji, S., Sheng, M., Zhou, D., et al.: Flexible and distributed mobility management for integrated terrestrial-satellite networks: challenges, architectures, and approaches. IEEE Netw. **35**(4), 73–81 (2021)
7. Chen, Q., Giambene, G., Yang, L., Fan, C., Chen, X.: Analysis of Inter-Satellite link paths for LEO mega-constellation networks. IEEE Trans. Veh. Technol. **70**(3), 2743–2755 (2021)
8. Handley, M.: Delay is not an option: low latency routing in space. In: Proceedings of the 17th ACM Workshop on Hot Topics in Networks, New York, pp. 85–91 (2018)
9. Werner, M.: A dynamic routing concept for ATM-based satellite personal communication networks. IEEE J. Sel. Areas Commun. **15**(8), 1636–1648 (1997)
10. Ekici, E., Akyildiz, I.F., Bender, M.D.: A distributed routing algorithm for datagram traffic in LEO satellite networks. IEEE/ACM Trans. Netw. **9**(2), 137–147 (2001)
11. Lan, T., Li, H., Wu, Q., et al.: Exploiting path diversity to increase system performance in mega constellations. In: IEEE Wireless Communications and Networking Conference, Nanjing, WCNC, pp. 1–7. IEEE (2021)
12. Chang, H.S.: FSA-based link assignment and routing in low-earth orbit satellite networks. IEEE Trans. Veh. Technol. **47**(3), 1037–1048 (1998). https://doi.org/10.1109/25.704858
13. Taleb, T., Mashimo, D., Jamalipour, A., Kato, N., Nemoto, Y.: Explicit load balancing technique for NGEO satellite IP networks with on-board processing capabilities. IEEE/ACM Trans. Netw. **17**(1), 281–293 (2009)
14. Bai, J., Lu, X., Lu, Z., Wei, P.: Compact explicit multi-path routing for LEO satellite networks. In: Workshop on High Performance Switching and Routing, HPSR, Hong Kong, pp. 386–390 (2005)
15. Li, D., Mao, X., Yu, J., Wang, G.: A destruction-resistant dynamic routing algorithm for LEO/MEO satellite networks. In: The Fourth International Conference on Computer and Information Technology, CIT, Wuhan, pp. 522–527 (2004)
16. Li, D., Wang, X., Meng, Y.: A destruction-resistant routing algorithm in low earth orbit satellite networks. In: International Conference on Wireless Communications, Networking and Mobile Computing, Shanghai, pp. 1841–1844 (2007)
17. Han, C., Liu, A., Huo, L., Wang, H., Liang, X.: Anti-jamming routing for Internet of satellites: a reinforcement learning approach. In: IEEE International Conference on Acoustics, Speech and Signal Processing, ICASSP, Barcelona, pp. 2877–2881. IEEE (2020)

Research on Statistical Priority-Based Multiple Access Protocol in Mobile Ad Hoc Networks

Han Li[1], Shouming Wei[1](✉), Cheng Zhang[2], and Shuo Shi[1]

[1] Harbin Institute of Technology, Harbin 150001, Heilongjiang, China
weishouming@hit.edu.cn

[2] Aircraft Design Research Institute in Shenyang of Aviation Industry, Shenyang 110000, Liaoning, China

Abstract. Recently, flying Ad Hoc networks (FANETs) have been widely applied to various civil and military applications. Medium access control (MAC) protocols have attracted much attention due to its impact on the overall network performance. However, due to the unique characteristics of FANETs applied in military such as multi-priority differentiation requirements, high mobility and dynamic topology, traditional MAC protocols cannot meet the requirements of FANETs. Tactical targeting network technology (TTNT) proposes statistical priority-based multiple access (SPMA) that can support multiple service prioritization and provide extremely low access latency for high priority services. Therefore, the design and optimization for SPMA protocol has become a new research focus. This paper firstly compares SPMA protocol with carrier sense multiple access (CSMA) protocol, and introduces the working mechanism of SPMA protocol. Subsequently, the optimization of SPMA protocol is discussed and qualitatively compared from the perspectives of implementation methods, advantages and limitations. In the end, the issues and research challenges are summarized and discussed.

Keywords: SPMA protocol · FANETs · Algorithm Optimization

1 Introduction

In recent years, unmanned aerial vehicles (UAVs) have gained widespread use in academic research, commercial endeavors, and military applications owing to their cost-effectiveness, high mobility, stealth capability, self-organization, adaptability, and scalability [1, 2]. Examples of UAV applications encompass emergency communication [3], disaster search and rescue operations [4, 5], farmland monitoring [6, 7], post-disaster operations [8, 9], and traffic surveillance [10]. Utilizing multiple UAVs to collaborate rather than a single UAV has emerged as a focal point of current research due to its potential for enhanced efficiency, speed, and flexibility.

Flying Ad Hoc networks (FANETs) expand the concept of mobile Ad Hoc networks into the realm of the sky, enabling dependable and real-time network communication among multiple UAVs. This facilitates the swift and collaborative completions of various

X. Li et al. (Eds.): 6GN 2024, LNICST 615, pp. 55–68, 2025.
https://doi.org/10.1007/978-3-031-93193-2_5

tasks by multiple UAVs (see Fig. 1). FANETs primarily find utility in military applications for transmitting control commands and posture messages, including voices, images, texts, and videos [11]. Control commands necessitate low latency and a high success delivery rate to ensure precise communications, while other messages such as voice and video demand ample throughput and timeliness. The primary aim of MAC protocols is to manage packet transmission conflicts, minimize channel access delays, and facilitate equitable sharing of a common medium among multiple nodes. The MAC layer plays a pivotal role within FANETs, given their diverse array of multi-priority differentiated requirements, high mobility, and dynamic topology. Thus, when designing MAC protocols, it is essential to prioritize the classification of business information to meet the quality of service (QoS) requirements of different services, ensuring that high-priority business data experience lower latency and higher successful delivery rates.

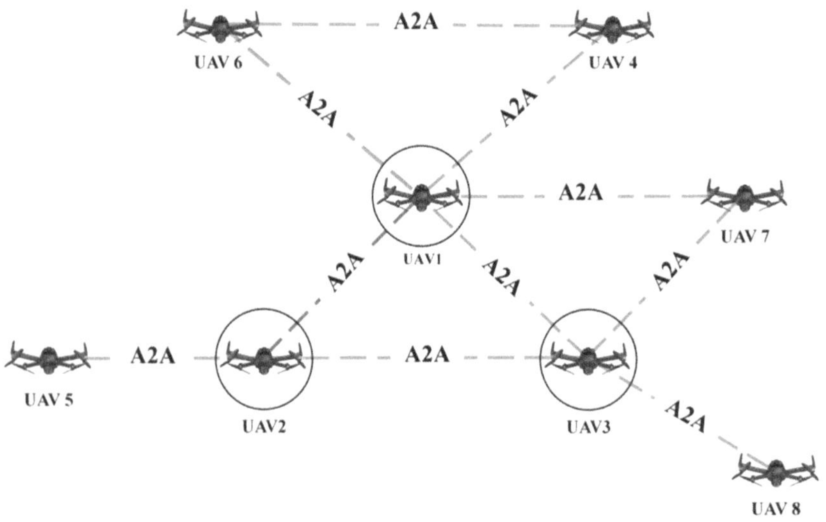

Fig. 1. Network architecture of FANETs.

Traditional MAC protocols, whether fixed allocation or randomized competition-based, fail to meet the requirements of FANET [12]. On one hand, fixed distribution protocols like time division multiple access (TDMA), frequency division multiple access (FDMA), code-division multiple access (CDMA), etc., offer high stability and ensure fair data transmission. However, their transmission delays are excessively long, making them unsuitable for highly dynamic network topology environments. On the other hand, random contention protocols such as carrier sense multiple access (CSMA) and carrier sense multiple access with collision avoidance (CSMA/CA) have relatively straightforward protocol mechanisms. Nevertheless, their collision probability rises substantially, especially in large network sizes, leading to severe performance degradation with increased traffic. Statistical priority-based multiple access (SPMA) serves as the MAC layer protocol for the U.S. Army's tactical targeting network technology (TTNT). It addresses the diverse service needs within the network and achieves a delay of less than

2 ms for the highest-priority service, with a successful delivery rate exceeding 99%. Consequently, there is an extensive literature on optimizing SPMA to design multi-priority MAC protocols for UAV ad hoc networks [13].

After conducting an extensive review of related literature, it was found that few scholars have summarized research in the direction of SPMA. This study aims to focus on, analyze, and summarize the research findings of numerous scholars from the perspective of mechanism optimization. Additionally, detailed descriptions of the basic principles of SPMA and the service access flow are provided. The primary contributions of this study are as follows:

- Differences between SPMA and CSMA protocol, operational mechanism of SPMA are addressed.
- According to the operational mechanism of SPMA protocol, existing optimization algorithms based on SPMA protocol are classified. Mechanism optimizations of SPMA are extensively reviewed and discussed in terms of implementation algorithms, advantages and limitations. Qualitative comparison of various types of Mechanism Optimizations of SPMA are qualitatively compared and discussed.
- Finally, research challenges are summarized and discussed for further improvement and development.

2 Mechanism Design of SPMA

In this section, CSMA and SPMA are initially compared and analyzed, with a focus on contrasting their channel state determination and back-off mechanisms. Subsequently, the operational mechanism of the SPMA protocol is introduced.

2.1 Comparison Between CSMA and SPMA

Prior to transmitting a service over the channel, the channel condition undergoes evaluation. However, unlike CSMA protocols, which categorize channel occupancy as simply 'busy' or 'idle', the SPMA protocol quantitatively assesses network congestion by tallying the number of services present in the network over a defined period of time. The results of the channel load statistics are periodically updated during network operations (see Fig. 2). When a node in the network requires sending a service, it evaluates the service priority threshold against the channel load statistics to determine whether to proceed with access or initiate a back-off.

Both the SPMA protocol and CSMA-like protocols implement a back-off mechanism following channel state evaluation. However, the SPMA protocol, designed to support multiple priority packets, employs a fundamentally different back-off algorithm. In CSMA/CA, the binary exponential back-off (BEB) Algorithm is utilized. This algorithm operates on the principle of doubling the back-off time after a failed attempt to access the channel, with subsequent attempts made after the back-off period elapses [14]. In contrast, the SPMA protocol necessitates service differentiation for various types of services, particularly prioritizing high-priority services to ensure better QoS. Unlike CSMA/CA, which defaults to the same service type for each node, the SPMA protocol may encounter situations where high-priority services are scheduled after low-priority services if the BEB algorithm is employed. Moreover, when the BEB algorithm employs passive back-off after conflicts have arisen, the SPMA protocol requires the back-off algorithm to autonomously determine whether to access the channel based on its occupancy status. Additionally, the back-off time in the BEB algorithm exponentially increases to reduce channel conflicts in networks with a high service load. However, this approach results in significant data transmission delays, which fail to meet the stringent delay requirements of FANETs in military-oriented communications.

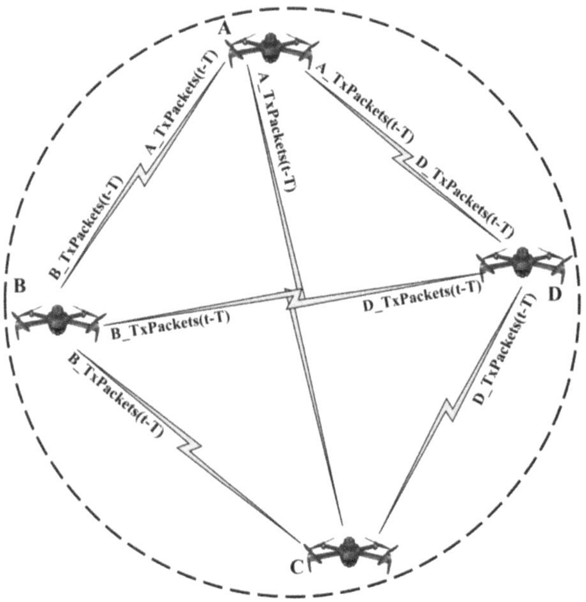

Fig. 2. Schematic of SPMA protocol channel occupancy statistics.

2.2 Operational Mechanism of SPMA

The state transfer process of SPMA protocol (see Fig. 3) mainly consists of channel detection process and multi-priority transmission decision process [15]. Initially, the network remains in its initial state unless a service arrives. Upon arrival, services from the application layer are queued up based on their priority, with high-priority services given precedence. Each queue is traversed according to the first-in-first-out principle (see Fig. 4).

Subsequently, each network member periodically updates the channel load statistics. When a service is scheduled for transmission, it is first checked for expiration. If expired, the service is removed from the queue and the queue is re-evaluated. Otherwise, the threshold value corresponding to the service's priority is compared with the channel occupancy statistic result. If the former is greater, the service is transmitted immediately.

However, if the latter is greater, the service undergoes a predefined back-off time before transmission.

Specifically, if a higher-priority service arrives during the back-off process, it takes precedence. Upon completion of the back-off process, each service is re-evaluated to ascertain if a timeout has occurred. If a timeout is detected, the service is removed; otherwise, the corresponding threshold value is compared with the channel occupancy statistic to determine whether transmission is permissible.

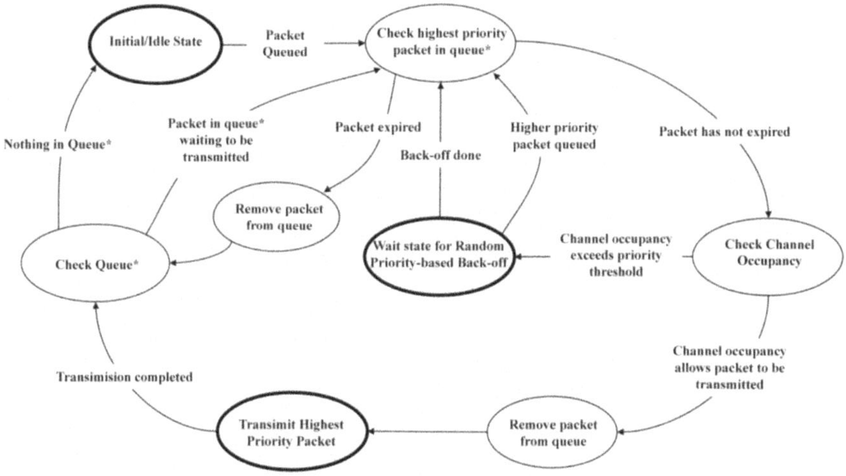

Fig. 3. The state transfer process of SPMA protocols.

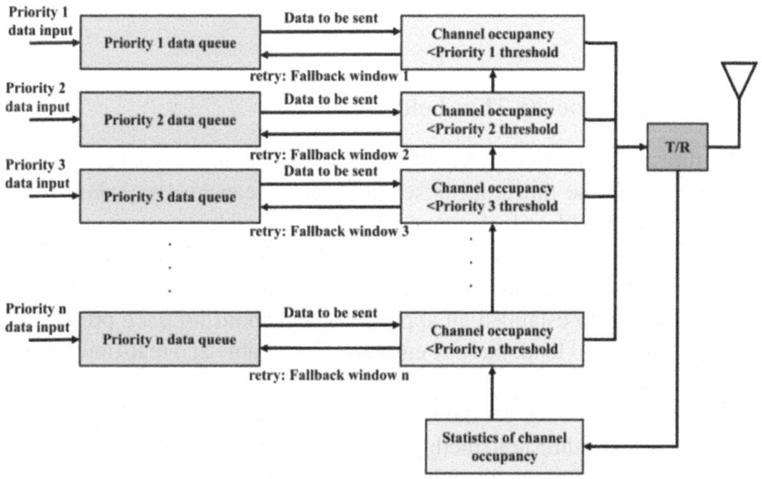

Fig. 4. Multi-priority transmission decision mechanism.

3 Parameters Optimization of SPMA

In this section, following a thorough comparison and analysis of the research findings regarding SPMA parameter optimization, we categorize them into three main aspects: priority threshold, channel occupancy statistics, and back-off time (see Fig. 5). Additionally, we delve into their respective optimization algorithms, advantages, and limitations to facilitate a comprehensive comparison.

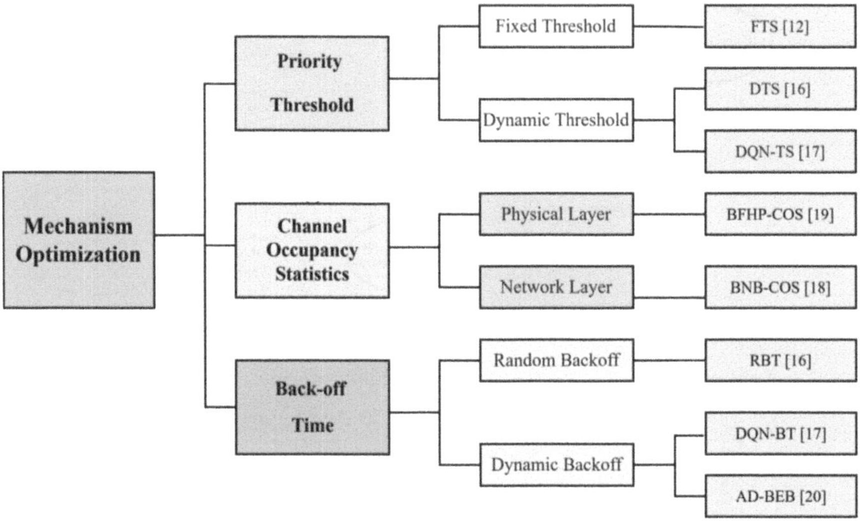

Fig. 5. The classification of parameters optimization.

3.1 Optimization of Priority Threshold

The priority threshold algorithm directly determines whether a packet can be immediately transmitted. If the threshold is set too high, it can lead to excessive channel load, increasing the likelihood of channel collisions and compromising the reliable transmission of high-priority data. Conversely, setting the threshold too low may excessively suppress data transmission, resulting in unnecessary wastage of channel resources. Therefore, it is crucial to establish an appropriate threshold for each priority queue to precisely regulate channel load, thereby ensuring the reliable and real-time transmission of high-priority data [12].

The paper [12] proposes to get the lowest priority threshold T_{N-1} by testing the data link layer. Next, the fixed threshold method (FTS) is utilized to estimate the thresholds for other higher priority levels. Initially, it is ensured that all test nodes are reachable within a single hop using the ALOHA mode of operation, where the channel load gradually increases with data transmission. When the success rate of the first transmission decreases to 99%, at this point the channel load value is recorded as the lowest priority threshold

T_{N-1} and the other higher priority thresholds are recursively represented as:

$$T_i = T_{N-1} * \frac{\sum_{j=0}^{N-1} r_j}{\sum_{j=0}^{i} r_j}, i \in (0, N-2), \tag{1}$$

where r_j is the proportion of each prioritized service volume in the total service volume of the network. The algorithm can increase the transmission success rate of high priority services and improve channel utilization. Nevertheless, the static threshold renders it incapable of effectively addressing unforeseen circumstances or frequently changing situations. Consequently, this results in a severe deterioration of network performance when the system load escalates.

Bujari A et al. [16] proposed a dynamic priority threshold (DTS) method based on the current priority level in the network. First, the threshold corresponding to the service flow with the highest priority in the current network is set to 1.0, and then it is determined whether only the lowest priority service flow exists in the network. If the answer is yes, then $T_{N-1} = 1.0$, if the answer is no, then $T_{N-1} = T$. Furthermore, it verifies whether the priority of this service flow is the next lowest priority. If affirmative, the computation terminates; otherwise, it proceeds to calculate the priority threshold for the corresponding priority level using Eq. (1). While the algorithm excels at adapting to changes in network traffic over time, it solely relies on priority levels to establish the priority threshold, neglecting other factors such as channel occupancy. This limitation poses challenges in managing the complexity of service traffic in large-scale networks.

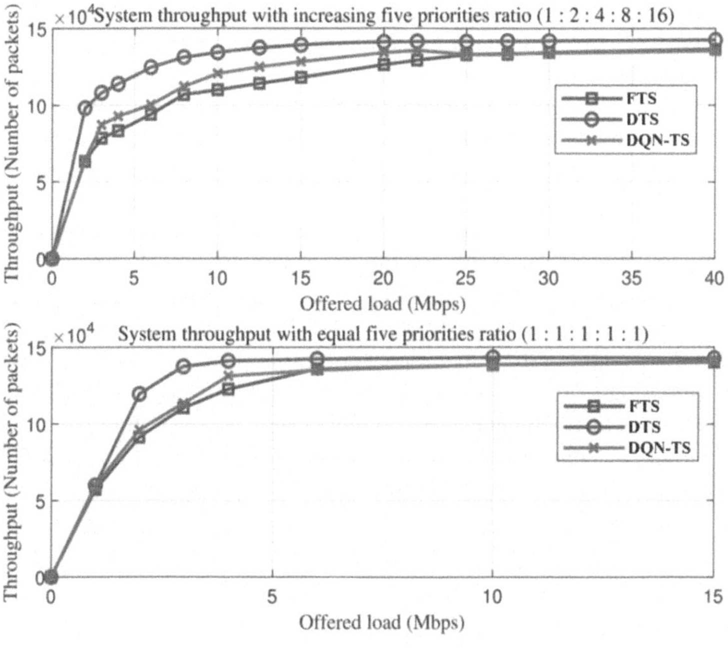

Fig. 6. Throughput of the network [17].

The paper [17] introduces a method called DQN-based priority threshold setting (DQN-TS) to enhance channel utilization for low-priority traffic, which greatly improves the throughput of the network (see Fig. 6). This method assesses whether a node can transmit a non-highest priority packet based on factors such as channel occupancy, the number of packets in each priority queue, and the arrival interval of each priority packet. At each time slot t, if a non-highest priority packet is queued for transmission and the current channel occupancy prohibits its transmission, the node employs DQN-TS to determine whether it can transmit the packet. The state space s_t:

$$s_t = \left[c_t, c_{t-1}, c_{t-2}, l_{0t}, \cdots, l_{(M-1)t}, i_{0t}, \cdots, i_{(M-1)t}\right], \tag{2}$$

includes the channel occupancy c_t and its previous values, the number of packets l_{jt} in each priority queue, the last arrival time of each priority packet i_{jt}, where j denotes the j_{th} priority. Then, by observing the state s_t from the state space, once the action is taken, the state of the network goes to s_{t+1} and the agent gets a reward r_t:

$$r_t = \begin{cases} 1, & \sum_0^{p-1} q_j = 0 \ or \ h_{p-1} - c_t \geq 1 \\ h_{p-1} - c_t, & \sum_0^{p-1} q_j \neq 0 \ and \ 0 \leq h_{p-1} - c_t \leq 1 \\ -1, & \sum_0^{p-1} q_j \neq 0 \ and \ h_{p-1} - c_t < 0 \end{cases}, \tag{3}$$

where h_j is the threshold of the j_{th} priority queue and q_j is the number of packets in the j_{th} priority queue. The algorithm takes into consideration a greater number of factors, while concurrently leveraging deep learning techniques to establish the priority threshold. This enables the transmission of additional low-priority packets without compromising high-priority performance, thereby enhancing communication network throughput. However, this enhancement comes at the cost of increasing computational complexity. Additionally, when confronted with new network environments, it demands substantial time and resources for training, along with ongoing model updates to adapt to network changes.

The fixed priority threshold setting method outlined in [12] offers simpler and lower algorithmic complexity. However, it suffers from poor flexibility and adaptability. This can lead to scenarios where low-priority services, which should ideally be restricted from accessing the channel, are granted access, resulting in heightened channel conflicts and degraded quality of service across all priorities. Conversely, there's the risk of delaying low-priority services unnecessarily, leading to increasing delays and decreased channel utilization. Dynamic priority threshold setting methods, as depicted in [16, 17], offer superior adaptability to evolving network traffic patterns over time. By allowing for the transmission of more low-priority packets without compromising high-priority performance, these methods enhance communication network throughput. However, they also introduce heightened computational complexity. The following Table 1 gives the comparison of priority threshold optimization algorithms.

Table 1. Comparison of priority threshold optimization algorithms.

Algorithm	FTS [12]	DTS [16]	DQN-TS [17]
Type	Fixed	Dynamic	Dynamic
Method	Test	Test	Theory
Priority	Yes	Yes	Yes
Channel Occupancy	No	No	Yes
Number of Packets	Yes	Yes	Yes
Latest Arrival Interval of Packets	No	No	Yes
Number of Priority	4	8	5
Simulator	OPNET	OPNET	OPNET
Complexity	Low	Medium	High
Adaptability	Low	Medium	High

3.2 Optimization of Channel Occupancy Statistics

The accuracy of channel occupancy statistics directly influences a node's assessment of network congestion, thereby impacting overall network performance. Low channel occupancy statistics may prompt a surge in service accesses, leading to heightened conflicts, diminished service transmission success rates, and degraded network throughput. Conversely, high channel occupancy statistics can trigger a surge in service back-off, potentially resulting in timeouts and service disruptions. This scenario culminates in increased delays, reduced channel utilization, packet transmission rates, and overall network throughput. Hence, the channel occupancy statistics algorithm holds immense significance [18].

The paper [19] introduces a channel occupancy counting method based on frequency hopping patterns (BFHP-COS) at the physical layer. This method prioritizes simplicity and accuracy. It relies on carrier detection to identify each pulse individually and assesses the occupancy of each hourly cycle, subsequently tallying the channel load across all frequency hopping patterns. The algorithm systematically evaluates each time frame to determine its occupancy status (see Fig. 7). If a frame is occupied, the load value increases by 1. Subsequently, the percentage of occupied time frames relative to the total number of frames is calculated to derive the final channel occupancy statistic. Compared to traditional methods that perform channel occupancy statistics for each frequency point, this algorithm significantly reduces computational complexity by only conducting channel occupancy statistics on the corresponding pattern during frequency hopping intervals. However, it is worth noting that while this physical layer-based method excels at capturing non-conflicting pulses, it fails to account for pulses lost due to collisions. Consequently, this limitation may result in understated channel occupancy statistics and reduced accuracy.

Li et al. [18] introduces a node broadcast-based channel load counting method (BNB-COS) operating at the network layer. This method tallies network load by regularly exchanging local traffic data among nodes in a distributed network and employs a sliding

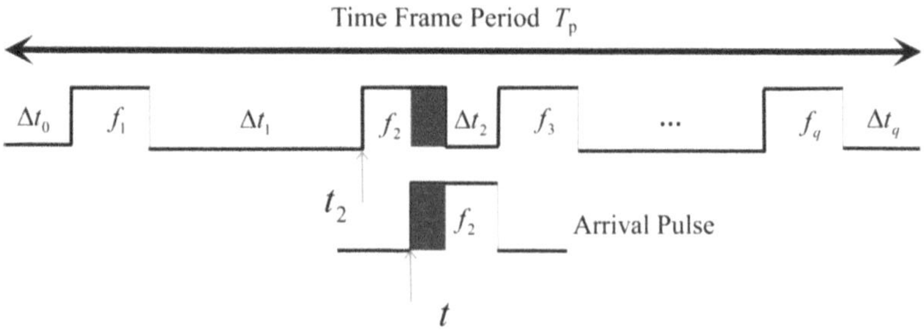

Fig. 7. Time frame occupancy diagram.

time window at the receiving end. Each node calculates channel load by tallying the total number of packets transmitted by all nodes and the total number of packets transmitted by the node itself within a time period T preceding a specific moment t, where T is the channel load counting period. Let the number of packets sent by a node i and all its neighboring nodes j in the last channel load statistics cycle be $TxPackets\_Local_i$, $RxPackets\_Neighbor_j$, respectively. Then node i's operations in the T is

$$Traffic_i(t - T) = TxPackets\_Local_i + \sum_{j=1,j \neq i}^{N} TxPackets\_Neighbor_j. \tag{4}$$

Let the number of nodes in the network be N and the length of the packet containing the channel load statistics be L. The estimate of the channel load statistics by node i at moment t is

$$Load(t) = Traffic_i(t - T) \times L/T = (TxPackets\_Local_i + \sum_{j=1,j \neq i}^{N} TxPackets\_Neighbor_j) \times L/T. \tag{5}$$

The algorithm leverages packets disseminated at the network layer to gauge channel load, enabling the counting of pulses lost due to collisions and thereby enhancing result accuracy. However, network layer broadcasting introduces extra delay and consumes channel bandwidth, diminishing the available bandwidth for data transmission patterns.

The channel occupancy statistical algorithm mentioned in [19] is characterized by low computational complexity. However, it only successfully detects pulses that are not in conflict and fails to account for pulses lost due to collisions, resulting in potentially inaccurate occupancy counting. Conversely, the node broadcast-based channel occupancy counting method is described in [18], operating at the network layer, significantly enhances counting accuracy. Nevertheless, it introduces additional delay and consumes channel bandwidth. The following Table 2 gives the comparison of channel occupancy statistics algorithms.

Table 2. Comparison of channel occupancy statistics Optimization Algorithms.

Algorithm	BFHP-COS [19]	BFHP-COS [18]
Type	Physical Layer	Network Layer
Node Broadcast	No	Yes
Number of Priority	4	4
Simulator	OMNeT++	OPNET
Latency	Low	High
Accuracy	Low	High

3.3 Optimization of Back-Off Time

The back-off algorithm addresses the challenge of determining the optimal back-off time to achieve efficient and reasonable data access [12]. If the back-off time is too short, it will not work to reduce the channel load. The back-off time is too long, and it will needlessly increase the access delay of data. In designing the back-off algorithm, it is imperative not only to devise an effective time differential generation mechanism and avoid excessively long back-off times, but also to ensure equitable allocation of channel resources among different priority access services.

The paper [16] proposes a method to randomize the back-off time (RBT) based on time slot size and packet priority. When the channel load is greater than its corresponding priority threshold, the back-off mechanism is performed and the back-off time is

$$t_{random\_backoff} = R(T, T * (p + 1)), \tag{6}$$

where T and p denote the time slot size and the priority of the packet. Moreover, the function $R(\cdot)$ denotes the generation of a random number on the interval $(\cdot)$. This approach offers simplicity in implementation, and the introduction of random back-off times can mitigate conflicts to some extent, thereby enhancing network throughput and efficiency. However, the drawback lies in the fixed setting of the back-off time, which prevents adjustment according to changes in the network state. Consequently, this limitation results in significant degradation of network performance when the network topology undergoes alterations.

Zhang et al. [17] introduced a method called DQN-based back-off time (DQN-BT) to minimize back-off durations. This algorithm calculates the back-off time by considering both packet status and channel occupancy. When the system transitions into the back-off state, it interacts with the network environment to gather the state information $C = \{c_t, c_{t-1}, c_{t-2}\}$, and subsequently computes the appropriate back-off duration. The expression for the back-off duration is as follows:

$$t_{backoff} = \frac{2a_t T}{N_s} + \frac{T}{2}, \tag{7}$$

where T is the length of the slot and N_s denotes the number of sub-pulses per packet. At the end of the back-off, the agent is rewarded r_t:

$$r_t = \begin{cases} 1, & B = 1 \; and \; t - t_s \leq T \\ \frac{1.25*T-0.25*(t-t_s)}{T}, & B = 1 \; and \; t - t_s > T \\ 0, & B = 0 \; and \; t - t_s \leq T \\ \frac{0.25*T-0.25*(t-t_s)}{T}, & B = 0 \; and \; t - t_s > T \end{cases}, \tag{8}$$

where t_s is the start time of the back-off. b is the back-off result, where 1 indicates successful back-off and 0 indicates failed back-off. The algorithm operates on a simple principle: the shorter the successful back-off time, the closer the reward approaches 1, on the contrary, the longer the failed back-off time, the closer the reward approaches -1. This method amalgamates q-learning and neural networks to analyze various input dimensions, such as channel occupancy and changes in channel occupancy. This comprehensive approach enhances the throughput of low-priority tasks within the network. However, it introduces heightened computational complexity and requires substantial time and resources for training in novel network environments. In the end, continuous updates of the model are required to adapt to network changes, which entail more time and human resources.

The paper [20] proposed an adaptive binary exponential back-off (AD-BEB) mechanism. The algorithm adaptively sets the minimum of the back-off window for different priorities:

$$W_0^{\xi P} = \left\lceil \frac{COS_\xi}{Th_\xi} W_0 \right\rceil, \xi = 1, 2, 3, \ldots, \tag{9}$$

where COS_ξ is the different channel load states detected by different priority nodes and Th_ξ denotes the threshold for different priorities. Moreover, ξP represents the initial competition value. If there is no successful transmission at the end of the back-off count, back-off until the maximum value is reached according to the following equation

$$W_i^{\xi P} = 2^{i-1} \cdot W_0^{\xi P}, i = 1, 2, 3, \ldots, m. \tag{10}$$

The algorithm not only enhances the message transmission success rate of high-priority nodes but also markedly diminishes the probability of system collisions and prevents conflicts arising from a large number of node messages. However, its simulation only incorporates three priority packets, which may compromise the efficacy and adaptability of the method in a broader network environment.

The algorithm outlined in [16] for randomly selecting the back-off time is relatively simple. While it does marginally reduce conflicts, leading to improved network throughput and efficiency, its adaptability is somewhat limited, resulting in only modest enhancements in overall network performance. On the other hand, the algorithm represented by [17, 20], which dynamically determines the back-off time, significantly diminishes the probability of system collisions and effectively mitigates conflicts stemming from a high volume of node messages. However, it comes at the cost of heightened computational complexity. The following Table 3 gives the comparison of back-off time optimization algorithms.

Table 3. Comparison of back-off time optimization algorithms.

Algorithm	RBT [16]	DQN-BT [17]	AD-BEB [20]
Type	Random	Dynamic	Dynamic
Priority	Yes	Yes	Yes
Time Slot	Yes	Yes	Yes
Channel Occupancy	No	Yes	No
Change in Channel Occupancy	No	Yes	No
Number of Priority	4	5	3
Simulator	OMNeT++	OPNET	OPNET
Complexity	Low	High	Low
Adaptability	Low	High	Medium

4 Conclusion

This study delves into the SPMA protocol for FANETs, focusing on its operational mechanisms and parameter optimization. Initially, through a comparative analysis with other MAC protocols, we demonstrate the superior suitability of the SPMA protocol for FANETs. Subsequently, we provide a comprehensive overview of the implementation methods, advantages, and limitations of optimization algorithms tailored for the SPMA protocol in FANETs scenarios. Finally, we summarize the persistent challenges and issues in SPMA protocol optimization, offering valuable insights for researchers in this area.

The SPMA protocol's superiority has sparked extensive inquiry into its operational mechanisms and parameter optimization within FANETs. However, there are still many research questions and challenges that need to be addressed. On one hand, the limited number of priorities selected for simulating emerging, promising optimization algorithms combined with deep learning hinders their effectiveness and adaptability across broader network contexts. This may be attributed to the extensive experimental databases required for optimization algorithms combined with deep learning, while the conditions for testing real-world data are more stringent. Therefore, alongside refining the optimization algorithms themselves, researchers should also focus on constructing a more expansive database of optimal SPMA protocol parameters to facilitate the practical application of these optimization techniques. On the other hand, achieving a balance between adaptability, accuracy, and computational efficiency eludes current optimization algorithms. Thus, there is a pressing need for researchers to develop adaptive SPMA protocols tailored to varying mobile speeds, communication requirements, and data types. Such protocols aim to maximize throughput and efficiency as much as possible, addressing the multifaceted challenges posed by diverse network environments.

Acknowledgement. This work has received support from the National Natural Science Foundation of China under Grant 62171158, National Key Laboratory of Air-based Information Perception and Fusion and Aeronautical Science Foundation of China under Grant 20230001004002 and the Fundamental Research Funds for the Central Universities under Grant 2242022k60006.

References

1. Jawhar, I., Mohamed, N.: Communication and networking of UAV-based systems: classification and associated architectures. J. Netw. Comput. Appl. **84**, 93–108 (2017)
2. Motlag, N.H., Taleb, T.: Low-altitude unmanned aerial vehicles-based Internet of Things services: comprehensive survey and future perspectives. IEEE Internet Things J. **3**(6), 899–922 (2016)
3. Arafat, M.Y., Moh, S.: Localization and clustering based on swarm intelligence in UAV networks for emergency communications. IEEE Access **6**, 59891–59906 (2018)
4. Arafat, M.Y., Moh, S.: Location-aided delay tolerant routing protocol in UAV networks for post-disaster operation. IEEE Internet Things J. **6**(5), 8958–8976 (2019)
5. Erdelj, M., Krol, M.: Wireless sensor networks and multi-UAV systems for natural disaster management. Comput. Netw. **124**(4), 72–86 (2017)
6. Tian, X.L.: Development of a low-cost agricultural remote sensing system based on an autonomous unmanned aerial vehicle (UAV). Biosys. Eng. **108**(2), 174–190 (2011)
7. Lottes, G., Nefzi, B.: UAV-based crop and weed classification for smart farming. In: IEEE International Conference on Robotics and Automation, ICRA 2017, pp. 3024–3031 (2017)
8. Tuna, N.H., Taleb, T.: Unmanned aerial vehicle-aided communications system for disaster recovery. J. Netw. Comput. Appl. **41**, 27–36 (2014)
9. Sanchez-Garcia, J., Reina, D.G.: A distributed PSO-based exploration algorithm for a UAV network assisting a disaster scenario. Futur. Gener. Comput. Syst. **90**, 129–148 (2019)
10. Bujari, A., Calafate, C.T.: Flying ad-hoc network application scenarios and mobility models. Int. J. Distrib. Sens. Netw. **13**(10) (2017)
11. Herder, J.C., Stevens, J.A.: Method and architecture for TTNT symbol rate scaling modes. U.S. Patent 7839900, 11–23 (2010)
12. Sun, F., Deng, Z.: A networking scheme for FANET basing on SPMA protocol. In: 2020 IEEE 6th International Conference on Computer and Communications, ICCC, pp. 182–187 (2020)
13. Liu, X.: Research on Multi-priority Multi-channel MAC Protocol for FANET. Harbin Institute of Technology (2022)
14. Kwak, B.J., Song, N.O.: Performance analysis of exponential backoff. IEEE/ACM Trans. Netw. **13**(2), 343–355 (2005)
15. Li, Z., Zhao, J.: The performance analysis of SPMA protocol in high speed mobile ad hoc network. IEEE Access **11**, 23485–23494 (2023)
16. Wang, Y.L.: Research on Key Technologies of MAC Protocol in TTNT Tactical Data Link. Beijing Jiaotong University (2018)
17. Zhang, X., Lei, M.: A transmission and backoff method based on deep reinforcement learning for statistical priority-based multiple access network. In: 2021 IEEE 94th Vehicular Technology Conference, VTC2021-Fall, pp. 1–5 (2021)
18. Li, Y.: Research on Multiple Access Control Strategy Based on Statistical Priority. Harbin Institute of Technology (2022)
19. Liu, Y.: Research on Statistical Priority Multiple Access Protocol for Network. Harbin Institute of Technology (2020)
20. Cheng, P., Yang, X.: Algorithm improvement and modeling analysis for SPMA. In: 2023 9th International Conference on Computer and Communications, ICCC, pp. 2301–2305 (2023)

Edge Computing and Resource Allocation

Deep Reinforcement Learning for Cache Allocation of Multi-cloud Content Providers in Mobile Edge-Cloud Computing Networks

Gang Wang[1], Jinlong Hao[1(✉)], Guozeng Xu[1], Jian Tang[1], Xiuhua Li[1], and Penghua Li[2]

[1] School of Big Data and Software Engineering, Chongqing University, Chongqing, China
{GangWang,x.guozeng,kt4ngw}@stu.cqu.edu.cn, haojl@changan.com.cn
[2] School of Big Data and Software Engineering, Chongqing University of Posts and Telecommunications, Chongqing, China
liph@cqupt.edu.cn

Abstract. With massive smart devices accessing the Internet, multi-tier computing has shown promise as a system-level network architecture by deploying caching capabilities at edge servers that are closer to the user. However, constrained by factors such as the dynamic nature of user content requests and limited cache resources, cache allocation is key to improving network performance. To address these challenges, we investigate and formulate the cache allocation problem for maximizing the cache hit ratio in mobile edge-cloud computing networks. We propose a framework to address this issue based on double deep Q-networks. The framework adaptively allocates cache slots to cloud content providers (CCPs) based on variations in user demand for different CCPs and temporal fluctuations in this demand. Simulation results depict a measurable enhancement in the cache hit ratio of the proposed framework compared to the baseline frameworks.

Keywords: Mobile Edge-Cloud Computing Networks · Edge caching · Cache allocation

1 Introduction

In recent years, data traffic has exploded with the surge of mobile devices accessing the Internet. Concurrently, the demand for low latency and high bandwidth in modern applications has posed significant challenges to network infrastructure [1,2]. To overcome such issues, multi-tier computing, which deploys computing and storage resources at the network edge close to users, has emerged as a prospective architecture [3,4]. In next-generation mobile networks, base stations equipped with servers can cache content near the users, thus decreasing latency

X. Li et al. (Eds.): 6GN 2024, LNICST 615, pp. 71–82, 2025.
https://doi.org/10.1007/978-3-031-93193-2_6

and easing the burden on the core network. This process is known as edge caching [5–7].

Despite these advantages of edge caching, however, compared to cloud servers, edge servers (ESs) typically have limited cache space. Additionally, because content requests of users are dynamically changing, this often leads to a high number of cache misses, meaning the requested content is not available on the ESs [8,9]. Therefore, efficient cache allocation is crucial for optimizing system performance and has become a key focus for many researchers.

The authors presented a strategy for joint service caching, cache allocation, and computation offloading in a three-tier collaborative mobile edge computing system to optimize latency and energy consumption in [10]. The research in [11] further considered the impact of user location on service demand, designed a location-aware caching policy, and employed deep learning techniques to predict the optimal caching decision, thereby enhancing system responsiveness. To address the distributed caching problem of popular services in converged networks, [12] proposed a deep Q-learning-based service scheduling policy aimed at optimizing service scheduling in a limited cache space to maximize equivalent throughput. Additionally, Sun et al. [13] proposed a framework for edge caching with decentralized recommendations and designed a discrete multi-agent reinforcement learning algorithm to optimize resource utilization and improve overall system performance in mobile edge-cloud computing. In multilayer computing networks, Wang et al. [14] investigated the large-scale multiple-input multiple-output-assisted joint task offloading and caching optimization problem, with the aim of minimizing the total task scheduling latency while taking into account channel state information (CSI) imperfections.

Although the aforementioned study provides detailed insights, they only consider a single service provider in their scenarios. However, ESs need to serve multiple cloud content providers (CCPs) in the real world. Each CCP has distinct content distribution requirements and user request patterns, necessitating flexible caching resource allocation by the ESs. Therefore, in this paper, we focus our work on scenarios involving multiple CCPs. The contributions of this article are briefly summarised below

- We consider a multi-cloud content providers scenario for edge caching in one mobile edge-cloud computing network, optimizing cache allocation by focusing on the differences in user demands for various cloud content providers and how these demands change over time.
- We propose a DDQN-based framework that intelligently assigns cache slots to each cloud content provider according to the status of per time slot in the considered network.
- We experimentally prove the usefulness of the proposed framework, and the results show a remarkable enhancement in the cache hit ratio.

The rest of the paper is organized as follows. Section 2 outlines the system model as well as formulates the optimization problem. The proposed framework is shown in Sect. 3, and the performance of the proposed framework is evaluated in Sect. 4. Finally, the conclusion is delivered in Sect. 5.

2 System Model and Problem Formulation

In this section, we present an introduction to mobile edge-cloud computing networks first, then depict cache partitioning, followed by a cost model, and finally formulate the problem.

2.1 Mobile Edge-Cloud Computing Networks

As shown in Fig. 1, we consider a scenario within the framework of a mobile edge-cloud computing network. The network comprises a single ES and a set of users in its coverage area (denoted as $\mathcal{U} = \{1, 2, \ldots, U\}$), as well as multiple CCPs (denoted as $\mathcal{C} = \{1, 2, \ldots, C\}$). The ES acts as a compute node that caches limited content from cloud servers via backhaul links and is typically deployed in a geographic location close to the users. We assume that the ES virtualizes its resources (cache in our model), which consist of K cache slots, and allocates them among the C CCPs. Each CCP has the autonomy to determine the specific objects to cache within their assigned space. In the considered mobile edge-cloud computing network, we pay attention to events consisting of a series of time slots $\mathcal{T} = \{1, 2, \ldots, t\}$. The users' content requests are random, denoted as $R_t = \{r_{t_1}, r_{t_2}, \ldots, r_{t_n}\}$, where $|R_t|$ denotes the number of user requests at moment t, these requests are randomly directed to one of the multiple CCPs, and it is worth noting that the content provided by different CCPs follows a Zipf distribution with varying parameters [15]. Given that some objects provided by CCPs are dynamically generated or updated in real-time (e.g., live streams or real-time data), not all objects are qualified to be stored in the cache slots. To represent this situation, we assume that each CCP is given a specific cacheability ζ, representing a probability of the content requested by the user being cacheable.

2.2 Cache Partitioning Model

Given the limited cache resources on the ES, it is essential to develop an appropriate cache allocation strategy based on the characteristics of the CCPs, request patterns, popularity distribution, and other factors mentioned in Sect. 2.1. We presume that the ES allocates a total of K cache slots to C CCPs. Each CCP stores its most frequently requested content within its assigned slots, with each slot capable of storing only one object. The allocation can be represented by a vector $\boldsymbol{\delta} = (\delta_1, \ldots, \delta_C)$, where each δ_c expresses the number of slots assigned to the c-th CCP, and $\sum_{c=1}^{C} \delta_c \leq K$. The comprehensive set of possible allocations is defined as follows

$$\boldsymbol{\Phi} \triangleq \left\{ \delta | \sum_{c=1}^{C} \delta_c \leq K, \delta_c \in \mathbb{Z}^+ \right\}. \tag{1}$$

It is important to note that when the c-th CCP is allocated δ_c cache slots, it stores its δ_c highest-demand objects within these slots. Simultaneously, time is partitioned into discrete time slots. The ES can adjust the cache allocation

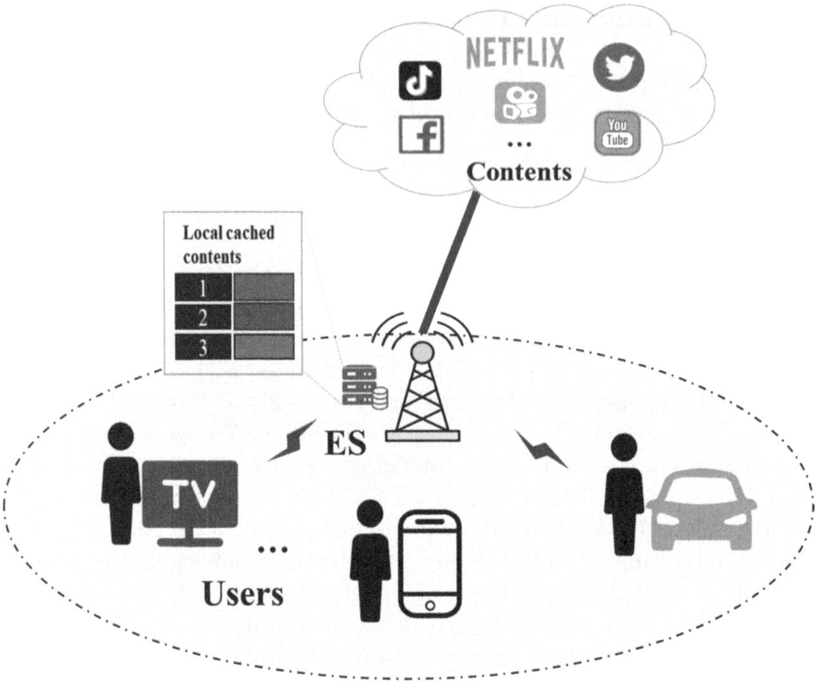

Fig. 1. Illustration of single ES and multiple CCPs for a mobile edge-cloud computing network.

for any given time slot. For instance, if Δ cache slots are added to a particular CCP, Δ cache slots will be subtracted from another CCP.

2.3 Cache Hit Model

We use $\rho_{t,r} \in \{0,1\}$ to indicate whether the user's request r at moment t is cached or not. Notably, in real systems, this term is both observable and recorded. The specific process is described as follows: once a user's request is directed to the CCP, the ES checks the cached content list to see whether the requested content is available locally. If the requested content is found in the cache, it is marked as 1, i.e., $\rho_{t,r} = 1$, indicating a cache hit. In this case, the ES directly delivers the content to the requesting user. Conversely, when the requested content is not in the cache, it is marked as 0, i.e., $\rho_{t,r} = 0$, indicating a cache miss. The ES then needs to download the uncached content from the remote server before delivering it to the user. This process generates uplink traffic, which can easily lead to congestion in the backbone network.

Based on the above analysis, we further define the proportion of direct cache hits relative to the total number of requests as the cache hit ratio. The calculation formula is as follows

$$\eta_t = \frac{\sum_r \rho_{t,r}}{|R_t|}, \tag{2}$$

where the interval of η_t is [0, 1].

The greater the cache hit ratio, then the more efficient the system performance in all aspects and the higher the quality of experience (QoE) for the user, and vice versa.

2.4 Problem Formulation

In this paper, due to the limited cache space on the ES and also the changing content preferences of users over time, it is a key issue to maximize the cache hit ratio by optimally allocating caches across multiple CCPs. Specifically, we aim to find the optimal allocation to maximize the cache hit ratio by dynamically adjusting the cache slots allocated to each CCP per time slot. The formulated problem is expressed as below

$$\max_{\boldsymbol{\Phi}} \mathbb{E}\left[\sum_{i=t}^{\infty} \eta_t\right] \tag{3a}$$

$$s.t. \sum_{c=1}^{C} \delta_c \leq K, \delta_c \in \mathbb{Z}^+ \tag{3b}$$

where $\mathbb{E}[\cdot]$ represents the expectation of the system's time-varying parameters, and the constraint in (3b) ensures that the cache allocation policy does not exceed the total number of available cache slots K while simultaneously ensuring efficient utilization of all cache slots.

3 Proposed Framework Design

In this section, we put forward a framework that is based on Double Deep Q-Networks (DDQN) for optimizing the cache hit ratio in the considered mobile edge-cloud computing network. The framework dynamically adjusts cache allocation to maximize the cache hit ratio.

3.1 Reasons for Using DDQN

The Q-learning, one of the mainstream reinforcement learning algorithms, finds an optimal action selection strategy by learning one Q-value function, i.e., $Q(s, a)$ that represents the expected reward for taking an action a in state s. The rule for updating Q-learning is as follows

$$Q(s_t, a_t) \Leftarrow Q(s_t, a_t) + \alpha\left(r_t + \gamma \max_a Q(s_{t+1}, a) - Q(s_t, a_t)\right), \tag{4}$$

where α represents the learning rate used to tune the update step during the Q-learning learning process. $\gamma \in [0, 1)$ indicates the discount factor and is used to modulate the algorithm's emphasis on future returns.

However, since Q-learning relies on tabular representations, it struggles to operate effectively in high-dimensional state spaces. To overcome this limitation, Q-learning is merged with deep neural networks in deep Q-networks (DQN) to approximate the Q-table of a large state, denoted as $Q(s, a; \theta)$. Furthermore, DQN introduces such techniques as experience replay and target network to stabilize training and improve convergence. Experience replay stores past experiences and samples them randomly to destroy correlations in the training data, while target networks reduce oscillations by providing stable Q-value targets. These improvements make DQN a powerful extension of Q-learning, capable of addressing more complex and high-dimensional reinforcement learning challenges [16]. Despite its advancements, DQN can suffer from the overestimation of Q-values due to using the maximum action value in target calculations, potentially destabilizing the learning process and leading to suboptimal strategies and unstable learning. To mitigate this issue, DDQN was developed as an enhancement to DQN [17]. DDQN separates the action selection and evaluation processes by employing two neural networks. One is a network that selects the action (action selection network), and the other is a network for evaluating the selected action's Q-value (target network). This segregation mitigates the inherent overestimation bias in DQN, resulting in more accurate Q-value estimates and improved learning stability. The update rule for DDQN is

$$y_t^{\mathrm{DDQN}} \Leftarrow r_t + \gamma Q(s_{t+1}, \arg\max_a Q(s_{t+1}, a; \theta); \theta^-). \tag{5}$$

Here, action selection is performed using an online network with parameter θ, while action evaluation is done using the target network with parameters θ^-. This separation effectively reduces the overestimation bias present in DQN.

3.2 The Proposed DDQN-Based Framework

A Markov Decision Process (MDP) is utilized to formulate the cache allocation issue. Based on this formulation, we then propose a DDQN-based cache allocation framework to optimize the cache hit ratio. Our aim is to find an optimal cache allocation policy that enables ESs, maximizing cache hit ratio by efficiently allocating cache space among multiple CCPs. Therefore, the problem is represented as one tuple (S, A, r), which is detailed as follows.

1) State Space: The set of possible environmental states that an agent can perceive at any given time point is the state space. The problem in the paper is a discrete state space that can be expressed as follows

$$\mathcal{S} = \left\{ \boldsymbol{\delta} = (\delta_1, \ldots, \delta_C) | \sum_{c=1}^{C} \delta_c \leq K \right\}. \tag{6}$$

2) Action Space: The set of all possible actions that an agent can fulfill in each state is the action space. The ES selects an action from the available action space when the system is in the $\boldsymbol{\delta}$ state.

$$\mathcal{A}_\delta = \{\mathbf{a} = \Delta \cdot (\mathbf{b}_c - \mathbf{b}_{c'}) | \delta + \mathbf{a} \in \mathcal{S}, c, c' = 1, \ldots, C\} \tag{7}$$

where $\mathbf{b}_c$ is an element of c-th of the $\mathbb{R}^c$ standard basis.

Algorithm 1 DDQN-Based Framework for Cache Allocation

Input: The state $\mathcal{S}$
Output: The optimal action $\mathcal{A}_\delta$

Initialize parameters α, γ, ϵ, and both evaluation θ and target Q networks θ^-, replay memory $\mathcal{D}$.
for each episode $e \in \{1, 2, \ldots, \mathcal{E}\}$ **do**
  Obtain the initial observation.
  for $t \in \{1, 2, \ldots, \mathcal{T}\}$ **do**
    Generate a random number $x \in [0, 1]$.
    if $x < \epsilon$ **then**
      Randomly select a_t,
    else
      Select the action a_t with the highest Q value from the target network.
    end if
    Perform the action a_t and obtain the reward r_t depending on (8) and observe one new state s_{t+1} in sequence.
    Store the observed tuple (s_t, a_t, r_t, s_{t+1}) into $\mathcal{D}$.
    Draw a mini-batch of experiences (s_j, a_j, r_j, s_{j+1}) from $\mathcal{D}$.
    Compute loss function $L(\theta)$ according to (10).
    Update θ;
    Update $\theta^- = \theta$ per Z steps.
  end for
end for

3) Reward Function: After each action, the reward function provides immediate feedback to the agent to help the agent assess the quality of the action. This feedback is typically represented by a real number, indicating the reward or penalty associated with performing an action at the current state. The aim of the agent is to maximize the cumulative total reward, often referred to as the discounted reward. In this paper, our target is to maximize the cache hit ratio. Therefore, we define r_t as follows

$$r_t = \eta_t. \tag{8}$$

The training process for DDQN begins with initializing the experience replay memory $\mathcal{D}$, which stores the agent's observed experiences. Additionally, the parameters of the two Q-networks are initialized: the online Q-network parameter θ and the target Q-network parameter θ^-. There is a ϵ-greedy strategy used in order to balance exploration and exploitation. To be specific, the agent chooses a random action to explore the environment with probability ϵ. Conversely, the agent chooses an action that corresponds to the highest Q value in the current

Q-network with probability 1 - ϵ in order to utilize the known information. After selecting an action a_t depending on the current state s_t, the agent performs a_t, obtains a reward r_t, and observes the next state s_{t+1}. This information is then formed into the tuple (s_t, a_t, r_t, s_{t+1}) and stored in the experience replay memory $\mathcal{D}$. Then a small batch of experiences (s_j, a_j, r_j, s_{j+1}) is randomly selected from $\mathcal{D}$ to update the network, calculated as follows

$$y_j \Leftarrow r_j + \gamma Q(s_{j+1}, \arg\max_a Q(s_{j+1}, a; \theta); \theta^-). \tag{9}$$

Updating the parameters of the evaluation Q-network θ by minimizing the loss function $L(\theta)$ is as follows

$$L(\theta) = \mathbb{E}_{(s_j, a_j, r_j, s_{j+1}) \sim D}[(y_j - Q(s_j, a_j; \theta))^2]. \tag{10}$$

Finally, at each Z-step, the evaluation Q-network θ is replicated in the target Q-network, θ^-, i.e., $\theta = \theta^-$, until the training stops after reaching a predefined number of training steps. Algorithm 1 provides detailed information about our proposed DDQN-based framework.

4 Simulation Evaluation

4.1 Simulation Settings

To evaluate the proposed dynamic cache allocation framework, we conduct experiments on a test platform running Ubuntu 18.04. The platform is equipped with 8 NVIDIA A800 GPUs. Unless otherwise specified, we consider a scenario involving three CCPs and one ES. The ES is assumed to have 200 cache slots available for caching popular content from the CCPs. User content requests are randomized and directed to CCP1, CCP2, or CCP3 with probabilities of 0.70, 0.20, and 0.10, respectively. Content provided by CCPs follows a Zipf distribution with varying parameters as 1.2, 0.6, and 0.8, respectively. We set $\alpha = 0.0001$ and $\gamma = 0.95$ in order to ensure better convergence of the proposed framework. Furthermore, the following three baseline frameworks were used to measure the validity of the proposed framework.

- **Random**: The ES randomly selects CCPs to increase or decrease the number of allocated slots.
- **Fixed**: The ES keeps the number of cache slots initially assigned to each CCP unchanged.
- **PRAP**: The ES allocates the cache slots based on the number of requests received from each CCP at the previous moment. Specifically, the more requests a CCP receives, the more slots are allocated to it, and vice versa.

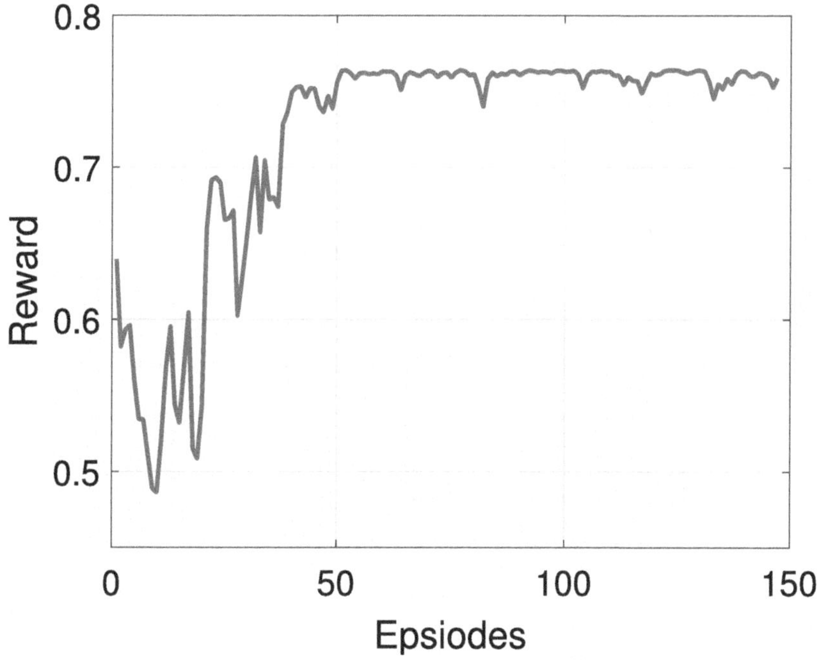

Fig. 2. The convergence curve of the proposed framework.

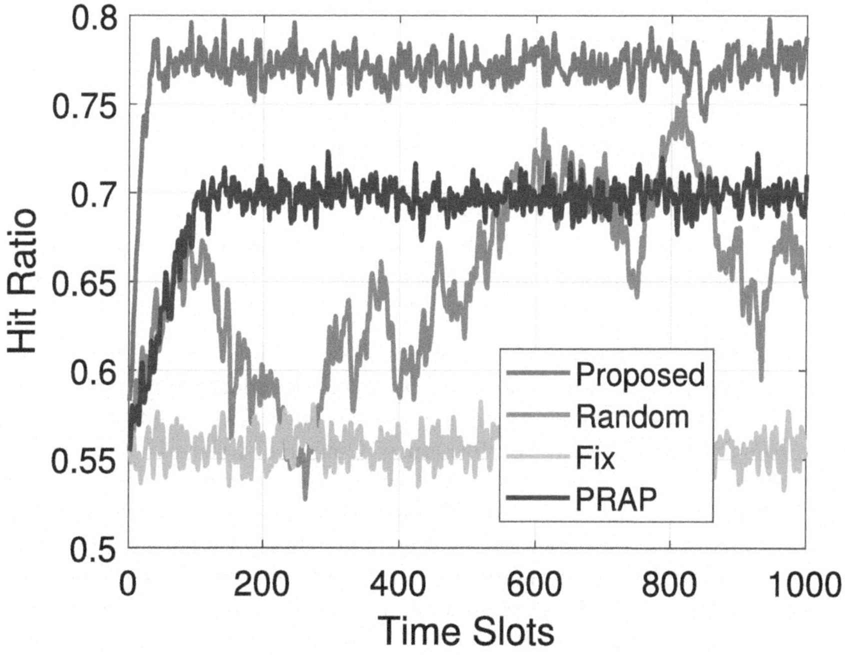

Fig. 3. The hit ratio of the proposed framework, Random, Fix, and PRAP.

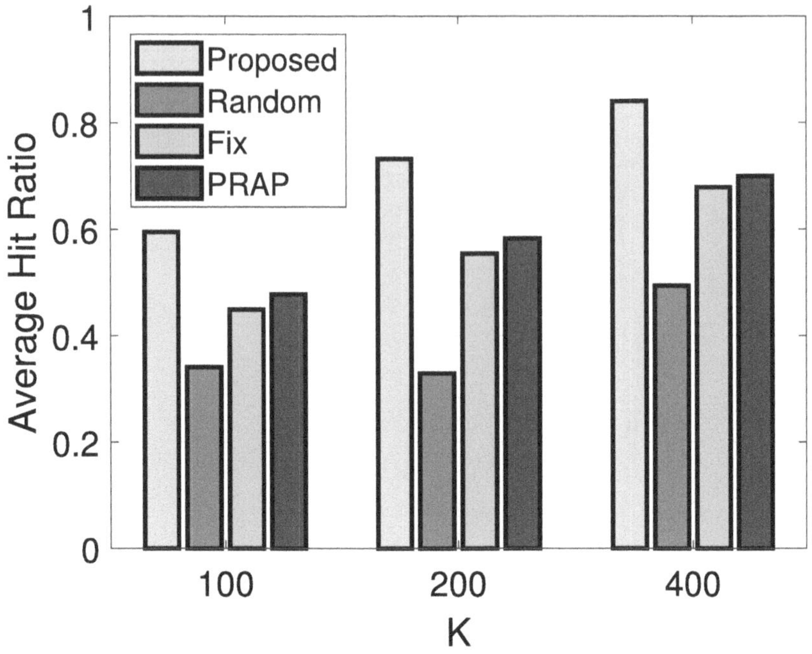

Fig. 4. Average hit ratio for different numbers of cache slots.

4.2 Simulation Results

1) Convergence of the Proposed Framework. The convergence curves of our proposed framework are revealed in Fig. 2. During the initial training phase, rewards fluctuate significantly as the agent explores various actions and refines its strategy. Around episode 50, rewards begin to stabilize, indicating effective utilization of learned strategies. By episode 150, the reward curve stabilizes near 0.8, demonstrating that the DDQN-based framework has successfully optimized cache allocation to maximize the cache hit ratio.

2) Comparisons of Hit Ratio. The hit ratio for the proposed framework, random, fixed, and PRAP, is presented in Fig. 3. The Random framework displays notable fluctuations throughout, highlighting its lack of inherent strategy and leading to suboptimal caching performance. Conversely, the Fixed framework, while more stable than the Random framework, consistently shows a lower hit ratio. This indicates that the Fixed framework is unable to adapt to varying content popularity and user access patterns. The PRAP framework outperforms the randomized and fixed methods but still fails to achieve the performance of the proposed DDQN strategy. It shows some stability but lacks the adaptive decision-making capability of the DDQN approach in dynamic environments.

3) Average Hit Ratio for Different Numbers of Cache Slots. Figure 4 displays the average hit ratio for different numbers of cache slots. As we can see, the cache hit ratio for all frameworks increases with the number of cache slots. Moreover, the proposed framework keeps the highest hit ratio compared to other frameworks. The random framework shows the lowest efficiency with significant fluctuations, while the fixed framework provides more consistent but still suboptimal results. Although the PRAP framework outperforms the random and fixed frameworks, they are still not comparable to the framework we proposed.

5 Conclusion

In this paper, we have investigated the issue of cache allocation in the considered mobile edge-cloud computing network. Specifically, we have considered multiple CCPs, then analyzed the cache hit model and formulated the problem of maximizing the cache hit ratio. To address the formulated problem, we have proposed a DDQN-based framework that intelligently allocates cache slots to each CCP based on the state of each time slot in the considered network. Simulation results have shown that the proposed framework dramatically improves the cache hit ratio compared to other baseline frameworks.

Acknowledgment. This work is supported in part by Science and Technology Innovation Key R&D Program of Chongqing (Grant No.CSTB2023TIAD-STX0035), National NSFC (Grants No. 62372072, 62102053, 62072060, 62067206 and 61972222), Chongqing Research Program of Basic Research and Frontier Technology (Grant No. cstc2022ycjhbgzxm0058), Haihe Lab of ITAI (Grant No. 22HHXCJC00002), the Natural Science Foundation of Chongqing, China (Grant No. CSTB2022NSCQ-MSX1104), Key Laboratory of Big Data Intelligent Computing, Chongqing University of Posts and Telecommunications (Grant No. BDIC-2023-B-003), Regional Innovation Cooperation Project of Sichuan Province (Grants No. 2023YFQ0028, 24QYCX0019), and Regional Science and Technology Innovation Cooperation Project of Chengdu City (Grant No. 2023-YF11-00023-HZ), and Fundamental Research Funds for the Central Universities (Grant No. 2024CDJGF-003).

References

1. Nguyen, D.C., et al.: 6G internet of things: a comprehensive survey. IEEE Internet Things J. **9**(1), 359–383 (2021)
2. Wang, X., Han, Y., Leung, V.C., Niyato, D., Yan, X., Chen, X.: Convergence of edge computing and deep learning: a comprehensive survey. IEEE Commun. Surv. Tutor. **22**(2), 869–904 (2020)
3. Jiang, N., Deng, Y., Nallanathan, A., Yuan, J.: A decoupled learning strategy for massive access optimization in cellular IoT networks. IEEE J. Sel. Areas Commun. **39**(3), 668–685 (2020)
4. Hoang, T.H., Nguyen, C.T., Do, T.N., Kaddoum, G.: Joint task offloading and radio resource management in stochastic MEC systems. IEEE Trans. Commun. (2024)

5. Cong, R., Zhao, Z., Min, G., Feng, C., Jiang, Y.: EdgeGo: a mobile resource-sharing framework for 6G edge computing in massive IoT systems. IEEE Internet Things J. **9**(16), 14521–14529 (2021)
6. Liu, B., Liu, C., Peng, M.: Resource allocation for energy-efficient MEC in noma-enabled massive IoT networks. IEEE J. Sel. Areas Commun. **39**(4), 1015–1027 (2020)
7. Liu, J., Ren, J., Zhang, Y., Peng, X., Zhang, Y., Yang, Y.: Efficient dependent task offloading for multiple applications in MEC-cloud system. IEEE Trans. Mob. Comput. **22**(4), 2147–2162 (2021)
8. Ben-Ameur, A., Araldo, A., Chahed, T.: Cache allocation in multi-tenant edge computing via online reinforcement learning. In: ICC 2022-IEEE International Conference on Communications, pp. 859–864. IEEE (2022)
9. Pervez, F., Sultana, A., Yang, C., Zhao, L.: Energy and latency efficient joint communication and computation optimization in a multi-UAV-assisted MEC network. IEEE Trans. Wirel. Commun. (2023)
10. Wang, L., Zhang, G.: Joint service caching, resource allocation and computation offloading in three-tier cooperative mobile edge computing system. IEEE Trans. Netw. Sci. Eng. **10**(6), 3343–3353 (2023)
11. Chen, J., Xing, H., Lin, X., Nallanathan, A., Bi, S.: Joint resource allocation and cache placement for location-aware multi-user mobile-edge computing. IEEE Internet Things J. **9**(24), 25698–25714 (2022)
12. Fang, Y., Xiong, J., Cheng, P., Zhang, W.: Distributed caching popular services by using deep Q-learning in converged networks. In: 2019 IEEE 90th Vehicular Technology Conference (VTC2019-Fall), pp. 1–5. IEEE (2019)
13. Sun, C., Li, X., Wen, J., Wang, X., Han, Z., Leung, V.C.: Federated deep reinforcement learning for recommendation-enabled edge caching in mobile edge-cloud computing networks. IEEE J. Sel. Areas Commun. **41**(3), 690–705 (2023)
14. Wang, K., Chen, W., Li, J., Yang, Y., Hanzo, L.: Joint task offloading and caching for massive MiMO-aided multi-tier computing networks. IEEE Trans. Commun. **70**(3), 1820–1833 (2022)
15. Goemans, M., Li, L.E., Mirrokni, V.S., Thottan, M.: Market sharing games applied to content distribution in ad-hoc networks. In: Proceedings of the 5th ACM International Symposium on Mobile Ad Hoc Networking and Computing, pp. 55–66 (2004)
16. Van Hasselt, H., Guez, A., Silver, D.: Deep reinforcement learning with double Q-learning. In: Proceedings of the AAAI Conference on Artificial Intelligence, vol. 30, no. 1 (2016)
17. Ding, Y., et al.: DDQN-based trajectory and resource optimization for UAV-Aided MEC secure communications. IEEE Trans. Veh. Technol. (2023)

HCTRL: Hierarchical Clustering Tree Reinforcement Learning Algorithm for Efficient Content Caching

Ruohan Shi[1], Zihong Ming[1], Qilin Fan[1(✉)], Xiuhua Li[1,2], Kai Wang[3], and Xiang Deng[1]

[1] School of Big Data & Software Engineering, Chongqing University, Chongqing 400044, China
{shiruohan,fanqilin}@cqu.edu.cn, mingzihong@stu.cqu.edu.cn

[2] Haihe Laboratory of Information Technology Application Innovation, Tianjin 300072, China

[3] School of Computer Science and Technology, Harbin Institute of Technology, Weihai 264209, China
dr.wangkai@hit.edu.cn

Abstract. Edge caching technology is crucial for improving user experience and reducing the load on origin servers by deploying multiple caching nodes at network edges. Given the limited capacity of these edge caches, managing the space effectively to accommodate dynamically changing user requests is paramount. Traditional optimization and heuristic algorithms, while sometimes offering viable solutions, generally lack adaptability to dynamic environments, leading to potentially suboptimal performance. To this end, we propose a hierarchical clustering tree reinforcement learning algorithm for efficient content caching. Specifically, to overcome the inefficiencies of traditional reinforcement learning in large action spaces, the long short-term memory network is employed to preprocess request content sets, organizing them into a balanced hierarchical clustering tree. Then, the reward function is designed to maximize the average hit rate of requested video content. Furthermore, we utilize a simple recurrent unit to build a state tracker that continuously gathers and records interaction trajectories. Experimental results on the real iQIYI dataset confirm the effectiveness of our proposed approach.

Keywords: Edge Caching · Reinforcement Learning · Hierarchical Clustering

1 Introduction

The development of the fifth generation (5G) networks, alongside the exponential increase in IoT device proliferation and a growing user demand for bandwidth-intensive applications such as video streaming, online gaming, and virtual reality, are driving a marked increase in video traffic volumes. According to a report by Cisco, it is projected that the global internet user base will expand from 3.9

X. Li et al. (Eds.): 6GN 2024, LNICST 615, pp. 83–97, 2025.
https://doi.org/10.1007/978-3-031-93193-2_7

billion in 2018 to an estimated 5.3 billion by 2023. On one hand, there is an escalating volume of video traffic coupled with rising demands for superior quality of experience (QoE). On the other hand, the increased demand for bandwidth and other resources, spurred by emerging technological trends, necessitates rapid scaling in the delivery of real-time streaming content both in terms of scale and capacity. These pressures not only challenge the existing infrastructures but also lead to congestion within content delivery networks, subsequently causing noticeable declines in user QoE.

Edge caching has emerged as a highly promising technique for optimizing network performance and enhancing resource utilization efficiency. By storing frequently accessed data resources in proximity to the local area, edge caching effectively reduces network latency, alleviates server load, and improves system scalability. However, the strategic allocation of content within the cache, constrained by its limited space, remains a vital issue to be tackled.

Conventional solutions fall short of comprehensively addressing the intricate issues related to the storage and replacement of edge content. To address this challenge, we introduce a novel algorithm that merges learning and heuristic approaches, termed as the hierarchical clustering tree reinforcement learning (HCTRL) algorithm for content caching. The action space is structured as a tree, forming the foundation for a cache recommendation algorithm grounded in deep reinforcement learning (DRL). Employing the principles of reinforcement learning, we extract recommended cache content from the action space and subsequently employ the least recently used (LRU) algorithm for storage. The primary objective of this endeavor is to maximize the average hit rate, thereby effectively mitigating the strain on network bandwidth resources. Our main contributions are summarized as follows:

- We propose a novel cache replacement scheme that integrates reinforcement learning with heuristic strategies for cache decision-making, aimed at maximizing long-term returns. To capture the dependencies of content requests, we employ simple recurrent unit (SRU) to collect the historical trajectories of request interactions as environmental state features.
- We extract temporal features of videos using long short-term memory (LSTM) neural networks, and construct a balanced hierarchical clustering tree to restructure the user action space based on these features. This restructuring represents cache selection as paths from the tree root to specific leaves, significantly reducing the time complexity during training and decision-making phases.
- We validate the effectiveness of our approach through experiments conducted on the real iQIYI video dataset. Experimental results demonstrate the superior performance of HCTRL.

The rest of the paper is organized as follows. In Sect. 2, we introduce the related works. Section 3 presents the system model and problem formulation. The details of our proposed approach are presented in Sect. 4. We compare the experimental performance and analyze the results in Sect. 5. Section 6 concludes the paper.

2 Related Work

In recent years, the cache replacement issue within edge computing has garnered significant scholarly and industrial interest. Depending on the optimization strategies employed to tackle the cache replacement issue, we categorize the research into the following three classes.

2.1 Heuristic-Based Optimization Algorithms

Heuristic algorithms utilize simplified decision logic or empirical rules to manage cache content, aiming to enhance cache hit rates and reduce system or network resource retrieval demands. Traditional heuristic algorithms such as LRU, least frequently used (LFU) [1], and first in first out (FIFO) [2] are widely utilized across various systems to manage and optimize cache resources. Despite their effectiveness in specific scenarios, these traditional heuristic strategies exhibit limitations in adapting to dynamic changes in content popularity and complex request patterns in modern network environments. Yasir *et al.* [3] introduced a cache scheme considering content popularity and user preferences in the mobile edge computing (MEC) field, utilizing content-based collaborative filtering to predict user preferences. Similarly, Srivastava *et al.* [4] introduced a weight-based greedy cache replacement algorithm, which integrates access-weighted parameters. Their study demonstrated that the algorithm outperforms traditional LRU and static re-reference interval prediction algorithms. While heuristic-based approaches are favored for their simplicity and low resource requirements, they encounter challenges in rapidly adapting to changes in access patterns, potentially resulting in decreased cache hit rates.

2.2 Mathematics-Based Optimization Algorithms

In cache optimization, mathematics-based algorithms use exact models to analyze complex systems, predict trends, and design optimal solutions. These methods employ detailed equations to guide decisions, thus enhancing cache efficiency and overall network performance. Typical mathematical models include linear programming models and queueing models. For instance, Li *et al.* [5] used a method based on the quadratic knapsack problem, semi-definite relaxation techniques, and constrained convex processes to effectively reduce the average transmission power. Si *et al.* [6] introduced a regret-free cache algorithm based on the online mirror descent strategy. Mehrizi *et al.* [7] introduced a Poisson regressor model based on Gaussian processes, combined with Bayesian methods and Monte Carlo Markov chains, significantly enhancing content request prediction accuracy in fixed-edge cache systems. While mathematical methods can significantly enhance single-node cache performance in specific scenarios, they face several challenges. Firstly, the high complexity and computational costs of these models hinder their ability to quickly adapt to changing traffic patterns, limiting their real-time capabilities and generalizability. Secondly, the complex parameter adjustment process lacks adaptability to environmental changes, complicating practical deployment and maintenance.

2.3 Learning-Based Optimization Algorithms

In recent years, learning-based optimization methods have gained significant attention in cache decision-making due to their ability to enhance cache efficiency and overall system performance by adapting to changes in access patterns and content popularity. These methods primarily utilize machine learning and deep learning techniques to optimize cache content replacement strategies by predicting future access demands. For example, Doan *et al.* [8] utilized a convolutional neural network (CNN) model to predict the popularity of unreleased videos and optimized cache decisions through expert-recommended prediction methods. Qiu *et al.* [9] proposed an innovative cache replacement algorithm named OA-Cache, combining time convolution networks and attention mechanisms to maximize cache hit rates through oracle approximation and achieving model lightweighting through knowledge distillation techniques. Zhong *et al.* [10] developed an edge node content caching framework based on deep reinforcement learning (DRL), employing the Wolpertinger architecture to optimize long-term cache hit rates without requiring prior knowledge of content popularity distribution. Wu *et al.* [11] developed a dynamic content update strategy based on DRL, simultaneously employing an algorithm integrating LSTM networks and external memory to adapt to changes in content popularity in wireless cache networks. Sadeghi *et al.* [12] introduced a reinforcement learning framework to optimize content caching strategies for 5G network small cell base stations, dynamically addressing changing costs and maximizing the efficiency of network edge storage capacity. However, learning-based optimization algorithms still face challenges such as high model complexity, large demands for training data, and issues with real-time applicability [13,14].

3 System Model

We consider the remote cloud server hosts a set of N video contents, denoted as $F = \{f_1, f_2, \cdots, f_N\}$. These contents are divided into uniformly sized segments for efficient management and transmission. Temporally, time is segmented into discrete intervals $T = \{1, 2, \cdots, t\}$, where the set of user requests received by the edge base station is represented as $Q = \{q_1, q_2, \cdots, q_t\}$. Within each interval t, user requests for file f are indicated by $q_t^f \in \{0, 1\}$, where $q_t^f = 1$ denotes an active request for file f, and $q_t^f = 0$ indicates no request. The maximum storage capacity of the base station is defined as M, with its cache inventory represented by $L = \{l_1, l_2, \cdots, l_M\}$. The binary variable y_t^f signifies the storage status of content f within the base station at time t, $y_t^f = 1$ indicates storage, while $y_t^f = 0$ indicates non-storage. The range of content that the base station can accommodate is constrained by its maximal storage capacity, as elaborated below (Fig. 1):

$$\sum_{f \in \mathcal{F}} y_t^f \leq M, \tag{1}$$

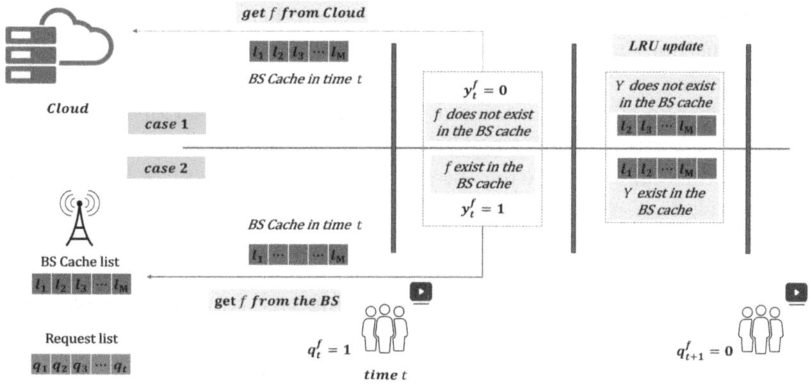

Fig. 1. System model.

At time t, if the requested content is already stored in the base station's cache, it can immediately deliver this content to the user, resulting in a cache hit, denoted by $q_t^f y_t^f = 1$. Conversely, if the base station does not have the required content in its cache, it must retrieve the content from the higher cloud layer to fulfill the user's request. We introduce H to represent the long-term average cache hit rate, calculated as T approaches infinity, and represented by the following equation:

$$H = \lim_{T \to \infty} \frac{1}{T} \sum_{f \in F} \sum_{t=1}^{T} q_t^f y_t^f. \tag{2}$$

After identifying the highly popular content Y as a potential hotspot, the classic LRU algorithm is employed to update and replace the cache content. If Y is already present in the cache, it is moved to the end of the cache list. If Y is not in the cache, the least recently accessed content c_1 is evicted, and Y is inserted at the end of the cache list, thereby updating the cache space.

4 Methodology

The RL algorithm learns the optimal strategy through interaction with the environment. In the RL algorithms, agents interact with the environment by executing actions and evaluate the quality of these actions based on the rewards returned by the environment. RL algorithms can employ neural networks to approximate the value function or policy function, thus enabling effective handling of large-scale state and action spaces.

4.1 MDP Formulation and Modeling

We introduce the formulation of a Markov decision process (MDP) to model the dynamic edge caching scenario under the RL framework. The selection strategy is designed to identify content with the highest predicted demand based on user requests, adjusting the policy iteratively in response to reward signals until the model stabilizes.

Specifically, the cloud maintains a comprehensive dataset of video contents. The model's task involves selecting the content that is most likely to be requested at a given time and facilitating pre-caching. This content selection mechanism is modeled as an MDP, utilizing policy gradient descent techniques to refine decision-making strategies that maximize long-term benefits. Moreover, to address the convergence challenges in large-scale action spaces inherent in RL, we devise a hierarchical balanced clustering tree to restructure the action space effectively.

State. The state s_t represents the interaction status at any given time step, serving as a crucial input for the decision-making network by conveying current system information. It includes vectors that capture the temporal dynamics of content requests. In our model, state features are extracted using the SRU, which optimizes the efficiency of temporal data processing.

Action. The action a_t denotes the decision made at time t within the interactive decision-making process. Action selects the content which is most likely to be requested at that moment.

Reward. The variable r_t denotes the feedback provided by the environment following the selection of action a_t by the policy at time t. This feedback value serves as a metric to evaluate the efficacy of the action selection. The reward is denoted by the long-term average hit rate H, as determined by Eq. 2. The overarching objective of optimizing the content selection strategy lies in maximizing the cumulative reward over the trajectory sequence.

Policy Network. The policy network $\pi_\theta = \pi_\theta \{a_t \mid \mathbf{s}_t\}$ is responsible for action selection based on the current state. We develop a fully connected neural network with a ReLU function at the output layer to serve as the policy network. This network takes states as input and outputs a probability distribution for selecting an action, mathematically described as:

$$\pi_\theta (a_t \mid s_t) = \text{ReLu} \left(\sigma \left(W_s^T s_t + b\right)\right), \tag{3}$$

where σ denotes a nonlinear activation function, while $W_s \in \mathbb{R}^{d_s \times d_a}$ and $b \in \mathbb{R}^{d_a}$ represent the weight matrix and bias values, respectively. These parameters are subject to iterative refinement during the training phase.

4.2 The HCTRL Algorithm

In RL-based models, exploring large discrete action spaces is often inefficient due to the decision-making time complexity, which scales with the action space size. This issue is exacerbated by the high number of video content requests. To solve this, we introduce a balanced hierarchical clustering tree, Tree T, which restructures the content action space for faster decision-making.

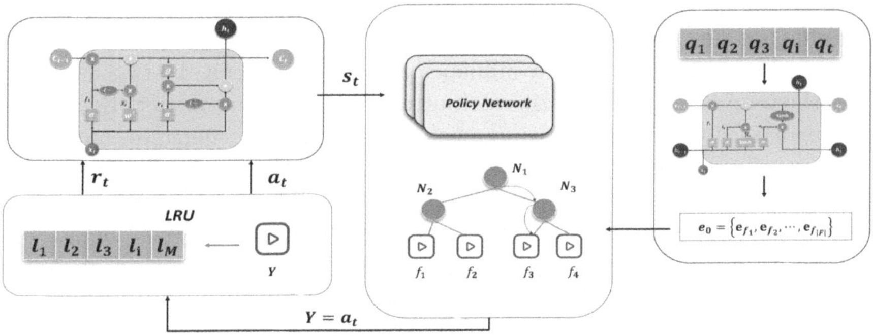

Fig. 2. The framework of HCTRL algorithm.

Figure 2 illustrates the framework of HTCRL algorithm, which consists of three critical components: feature extraction using LSTM to construct a hierarchical clustering tree, state extraction via the SRU model, and popularity prediction based on the tree structure. Initially, the LSTM algorithm preprocesses the data, followed by hierarchical clustering to organize the action space into a hierarchical tree, thereby streamlining the action space for RL. The SRU model then derives state vectors at each time step using historical data. These vectors serve as inputs to the RL framework, which predicts the most popular action at that time. Finally, the average hit rate is calculated to reflect the impact of the selected action on the overall caching system.

Construction of Hierarchical Clustering Tree. To facilitate more intuitive visualization and efficient implementation, we select a simple and widely adopted hierarchical method, namely the construction of a balanced tree structure. The initial phase of this process involves clustering all raw data points, specifically the embedding vectors of contents, into preliminary clusters. These clusters are progressively refined into sub-clusters until each content belongs to a single cluster. To maintain the balance of the tree structure, measures are taken to ensure that the height difference between any two subtrees of a node does not exceed one. Additionally, except at the final level, where adjustments might be necessary due to the limited number of contents, each non-leaf node maintains c children to preserve structural consistency and balance. The number of branches per node c is calculated based on the overall size of the content selection set F and the predetermined depth of the tree d, using a specific formula designed to ensure rational distribution and effective management across the tree. The formula for determining c is as follows:

$$c = \left\lceil N^{\frac{1}{d}} \right\rceil, \tag{4}$$

where N is the total number of contents, d is the depth of the tree and $\lceil x \rceil$ denotes the smallest integer not less than x, ensuring that each non-leaf node has enough children to partition the content space into manageable segments effectively. We use the LSTM algorithm to extract initial embedding vectors for video content due to its ability to understand the rich contexts and intents underlying request sequences. These sequences include complex variables like evolving user preferences, which influence behavior patterns. The LSTM model parses the temporal and semantic attributes embedded in users' historical behavior sequences. Its distinctive gating mechanism effectively manages information flow, capturing the long-range dependencies in sequential data. This capability is essential for understanding sustained user interactions and preferences over time. The processing of information within the LSTM framework includes the following components:

$$\begin{aligned} f_t &= \sigma\left(W_f \cdot \left[h_{t-1}, x_t\right] + b_f\right), \\ i_t &= \sigma\left(W_i \cdot \left[h_{t-1}, x_t\right] + b_i\right), \\ \widetilde{C}_t &= \tanh\left(W_C \cdot \left[h_{t-1}, x_t\right] + b_C\right), \\ C_t &= f_t * C_{t-1} + i_t * \widetilde{C}_t, \\ o_t &= \sigma\left(W_0 \cdot \left[h_{t-1}, x_t\right] + b_0\right), \\ h_t &= o_t * \tanh\left(C_t\right). \end{aligned} \tag{5}$$

In this LSTM network, x_t denotes the input at time t, specifically the sequence of embeddings corresponding to content requests over time. The parameters W_f, W_i, W_C, W_0 represent the weight matrices, while b_f, b_i, b, b_0 are the bias vectors associated with the respective gates within the LSTM architecture. The output vector from the hidden layer at time t is represented by h_t. Using the LSTM network, the initial feature vectors of contents, denoted as $\mathbf{e}_0 = \left\{\mathbf{e}_{f_1}, \mathbf{e}_{f_2}, \cdots, \mathbf{e}_{f_{|F|}}\right\}$, are obtained.

We employ principal component analysis (PCA) to perform dimensionality reduction on the initial vectors of content, denoted as $\mathbf{e}_0$. Following the dimensionality reduction, the feature data are subjected to hierarchical clustering to create a structured grouping. The clustering procedure begins with the establishment of an initial set of clusters. designated as O. This set $O = \left\{o_1, o_2, \ldots, o_{|F|}\right\}$ comprises individual clusters o_i, each initially containing a single feature vector $\mathbf{e}_{f_i}$. The process initiates with the application of PCA, which serves to standardize all feature vectors within o_i by reducing their dimensions. Subsequently, a covariance matrix is constructed to facilitate the clustering analysis.

$$\Sigma = \frac{1}{n-1}\left(o_i - \overline{o_i}\right)^T\left(o_i - \overline{o_i}\right), \tag{6}$$

where $\overline{o_i}$ is the average vector of o_i.

Then, we calculate the eigenvalues and eigenvectors of covariance matrix:

$$\Sigma V = \lambda V. \tag{7}$$

Eigenvalues are sorted according to their magnitude. The top k eigenvectors are selected to form the eigenvector matrix V. The vector o_i is then projected into the lower-dimensional space defined by these eigenvectors.

$$o_i' = V_k^T o_i. \tag{8}$$

The dimensionally reduced feature vectors are then processed using hierarchical clustering. Initially, distances are computed between the points within each cluster o_i'. The algorithm merges the two nearest points within the set o_i', forming a new cluster. This merging process continues iteratively until c distinct clusters are formed. This methodology takes a set of content vectors $\mathbf{e_0}$ and a specified number of child nodes c, partitioning these vectors into c clusters. Through multiple iterations of the PCA clustering algorithm, the clusters are refined until each cluster is uniquely associated with a specific content request.

Extraction of State Features. The SRU model is used to capture the state vector s_t. This state vector is specifically used to extract information on interaction scenarios that occurred before time t. It includes basic features of all requests, features of recently selected content vectors, and a sequence of actions taken during dynamic interactions, along with their corresponding rewards. Given the sequential decision-making nature of selecting video content, the SRU model effectively represents the state vectors dynamically. During the learning process for the t-th state vector $\mathbf{s}_t$, the model processes the sequence of previously selected target content embedding vectors $\{a_1, \cdots, a_n\}$ and their respective rewards $\{r_1, \ldots, r_{n-1}\}$ as inputs. These inputs enable the model to dynamically update the state h_t' by integrating the current input with historical state information. The operational flow of the SRU includes several key steps of information processing.

$$\begin{aligned}
\tilde{x}_t' &= W' x_t', \\
f_t' &= \sigma\left(W_f' x_t' + b_f'\right), \\
r_t' &= \sigma\left(W_r' x_t' + b_r'\right), \\
c_t' &= f_t' \odot c_{t-1}' + (1 - f_t') \odot \tilde{x}_t', \\
h_t' &= r_t' \odot g\left(c_t'\right) + (1 - r_t') \odot x_t',
\end{aligned} \tag{9}$$

here, x_t' represents the input at time t, which includes features about requests, actions, and rewards. $\tilde{x}_t'$ denotes a linear transformation, while W', W_f' and W_r' represent different weight matrices. b_f' and b_r' are the corresponding bias terms. c_t' and h_t' represent the cell state and output at time t, respectively. The symbol σ denotes the Sigmoid activation function. The notation $\odot$ signifies the Hadamard product.

Building A Reinforcement Learning Network Architecture. We use a balanced clustering tree model for video decision-making based on state s_t. Figure 2 shows the decision tree T, with a depth of 2, including three non-leaf nodes (N_1 through N_3). Each node is linked to a policy network

$(\pi(s_1), \pi(s_2), \pi(s_3))$ that makes decisions based on the current state s_t. The tree ends with four leaf nodes, each representing a different video content choice. The decision path P starts at the root node (N_1), where the state s is input into the policy network $\pi(s_1)$, directing the path toward the node with the highest probability (e.g., N_3) until reaching a leaf node (e.g., f_3). The traversal of P at time t is a sequence from the root to a chosen leaf node, represented by $\{c_1, c_2, \ldots, c_d\}$. Here, d is the path length, and c denotes the maximum number of child nodes per node. Determining a target content at time t involves navigating a path $P = \{c_1, c_2, \ldots, c_d\}$. Based on the state s_t, the probability of selecting action a_t is computed as follows:

$$\pi_\theta \left(a_t \mid s_t\right) = \prod_{i=1}^{d} \pi_{\theta_i} \left(c_i \mid s_t\right), \tag{10}$$

where $\pi_{\theta_i}\left(c_i \mid s_t\right)$ denotes the cumulative probability of decisions made by all policy networks from the root to the leaf nodes, as defined in Eq. 3. The model aims to optimize the entire set of policy networks represented by $\pi_\theta = \left\{\pi_{\theta_1}, \pi_{\theta_2}, \cdots, \pi_{\theta_Q}\right\}$, where Q indicates the count of non-leaf nodes within the tree, calculated according to the specified formula.

$$Q = \frac{c^d - 1}{c - 1}. \tag{11}$$

The policy networks $\pi_\theta = \left\{\pi_{\theta_1}, \pi_{\theta_2}, \cdots \pi_{\theta_Q}\right\}$ are trained to determine how each non-leaf node selects the optimal child node given a state s_t. The training employs the classic policy gradient descent method from RL. This method evaluates the policy of the model and utilizes the gradient information of the policy to update the model's parameters. The objective of HCTRL is to identify the most appropriate request content that maximizes the expected reward:

$$J\left(\pi_\theta\right) = \mathbb{E}_{\pi_\theta}\left[\sum_{t=0}^{n-1} \gamma^t r_t \left(s_t, a_t\right)\right], \tag{12}$$

where γ represents the discount factor, and r_t denotes the long-term average hit rate obtained at time t when a user request arrives. We can obtain the cumulative reward:

$$G^{\pi_\theta}\left(s_t, a_t\right) = \sum_{i=t}^{n} \gamma^{i-t} r_i, \tag{13}$$

herein, $\pi_\theta\left(a_t \mid s_t\right)$ represents the probability of taking action a_t given state s_t. Subsequently, parameters θ are updated using the gradient descent method:

$$\Delta\theta = \nabla_\theta \log \pi_\theta \left(a_t \mid \mathrm{s_t}\right) G^{\pi_\theta}\left(s_t, a_t\right), \tag{14}$$

$$\theta = \theta + \eta \Delta\theta. \tag{15}$$

The entire training process is detailed in Algorithm 1.

Algorithm 1: The HCTRL Algorithm

Calculate the number of child nodes c and the number of non-leaf nodes Q.
Initialize the environment and obtain the initial state vector s_0 according to the LSTM algorithm.
Based on s_0, construct a hierarchical balanced clustering tree T with c child nodes.
for $j = 1 \rightarrow Q$ **do**
  Randomly initialize the parameters θ_j.
end
for $t = 1 \rightarrow n$ **do**
  Compute the state s_t using SRU algorithm.
  Sample a path P using the tree structure.
  After the reinforcement learning algorithm, obtain the target selection content a_t.
  Calculate the reward value r_t.
  if $t < n$ **then**
    Calculate the state s_{t+1} at the next moment using the SRU model.
  end
  Get the trajectory M.
  for $t = 1 \rightarrow n$ **do**
    Update parameter θ via Eq. 14, Eq. 15.
  end
end

5 Performance Evaluation

We conduct the experiments based on the actual iQIYI dataset, which includes viewing data from 2 million users over two weeks and features 300,000 unique videos. For this study, we randomly select 10,000 content request records from the dataset, which includes timestamps, geographical locations, video IDs, and features related to video content. To simulate real-world caching scenarios, we assume that the initial storage is full and utilize a random storage policy. To study the effect of the suggested algorithm on cache hit rates, we set the following parameters: The balanced hierarchical clustering tree in the algorithm has a depth of 2, the discount factor is 0.995, and the initial learning rate is 0.005.

To demonstrate the effectiveness of our proposed HCTRL, we compare it with the following caching strategies: (1) LRU [1]: The LRU algorithm evicts the data that has not been accessed for the longest period. (2) LFU [1]: The LFU algorithm prioritizes eviction for the least frequently accessed data items. (3) FIFO [2]: The FIFO algorithm evicts data based on the order of entry into the cache, with the earliest cached data items being evicted first. (4) Lecar [15]: The Lecar strategy combines the advantages of LRU and LFU by dynamically adjusting the weight ratio of these two strategies to adapt to different data access patterns. (5) Arc [16]: The Adaptive Replacement Cache (Arc) strategy maintains two LRU lists and dynamically adjusts their size ratio to automatically adapt to actual access patterns. (6) Cacheus [17]: The Cacheus algorithm is an

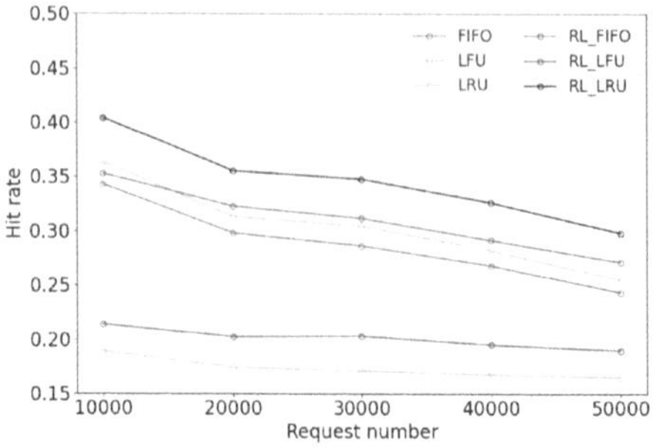

Fig. 3. The performance of different heuristic algorithms.

adaptive machine learning cache replacement strategy. It employs a set of experts (including ARC, LIRS, LFU, and two new experts, SR-LRU and CR-LFU) to handle various workloads by combining these experts for effective processing. (7) Belady [18]: Known as an optimal theoretical model within cache replacement strategies, Belady algorithm makes decisions based on future knowledge, evicting the content that will not be accessed for the longest time in the future. With a perfect understanding of the future, Belady algorithm can ensure the theoretical maximum cache hit rate. (8) DRL [14]: This strategy employs a reinforcement learning replacement algorithm based on the Deep Q-Network (DQN) algorithm. It learns through interaction with the environment to select the most appropriate content to evict from the cache list, thus implementing cache replacement.

We investigate the impact of three classical cache replacement algorithms (LRU, LFU, and FIFO) within a hierarchical clustering-based reinforcement learning caching strategy. The objective is to evaluate how different cache replacement policies affect system performance when the model predicts optimal cache contents. Experiments are conducted using datasets ranging from 10,000 to 50,000 entries to assess these strategies in practical applications. As depicted in Fig. 3, the LRU algorithm consistently outperforms LFU and FIFO across all scenarios, particularly when integrated with the reinforcement learning model. The effectiveness of LRU is attributed to its consideration of access frequency and data recency, thereby optimizing cache resources and improving hit rates. In contrast, LFU neglects data recency, leading to inefficiencies, while FIFO disregards access patterns, resulting in suboptimal performance in dynamic environments. By prioritizing the removal of the least recently accessed data, LRU enhances cache utilization and overall system performance.

Then, we examine the hit rate variation over content numbers. As shown in Fig. 4, cache hit rates for all algorithms decrease as the content number grows. This decline is due to the increased diversity of user requests, which reduces the likelihood of satisfying each request from the cache. With limited storage, an increase in total video requests decreases the probability of a cache hit. The proposed HCTRL algorithm outperforms other methods, except for the optimal Belady algorithm, improving average hit rates over the DRL, Arc, Lecar,

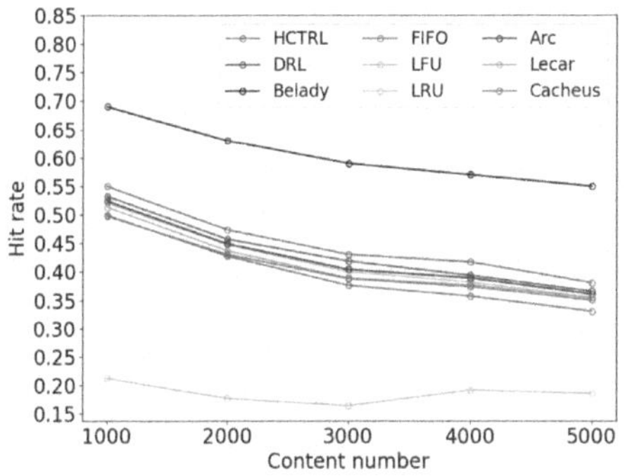

Fig. 4. The hit rates over different content numbers.

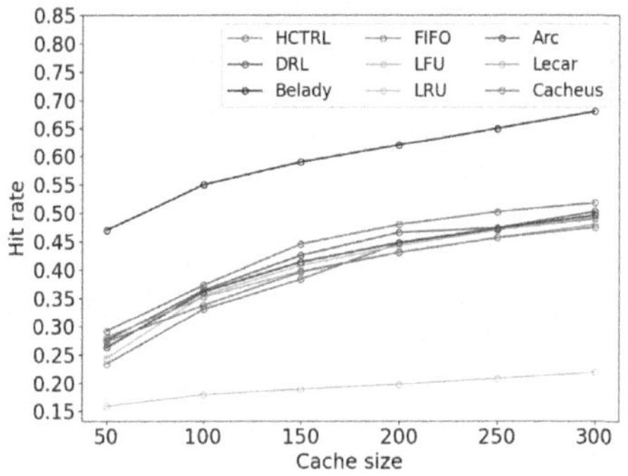

Fig. 5. The hit rate variation over different cache sizes.

Cacheus, LRU, FIFO, and LFU algorithms by 1.7%, 2.5%, 3.6%, 4.2%, 3.0%, 5.2%, and 26.4%, respectively.

In Fig. 5, we can see that as cache capacity increases, the hit rates of all caching algorithms increase. Higher hit rates mean more user requests are satisfied directly from the cache, reducing the need to access the source server, decreasing latency, and enhancing user experience. This improvement is due to the increased cache space, which allows for storing content that users are likely to access, directly impacting overall system performance. When comparing other caching algorithms except for the theoretically optimal Belady algorithm, the proposed HCTRL demonstrates superior performance. The HCTRL algorithm improves the average hit rate by 1.8%, 2.6%, 3.5%, 4.0%, 3.3%, 4.2%, and 23.9% over the DRL, Arc, Lecar, Cacheus, LRU, FIFO, and LFU algorithms, respectively, across different cache sizes. This indicates the efficiency of its optimization strategy for caching decisions. The performance advantage of the HCTRL algorithm arises from factors such as accurate prediction of user access patterns, rapid response to changes in content popularity, and optimal utilization of cache space.

6 Conclusion

In this paper, we have proposed a novel cache replacement scheme that integrates RL with heuristic strategy to optimize cache decision-making and maximize long-term returns. To capture content request dependencies, we have used SRU model to collect historical request trajectories as environmental state features. Temporal features of videos have been extracted using LSTM neural networks, and a balanced hierarchical clustering tree has been constructed to restructure the user action space, representing cache selection as paths from the tree root to specific leaves. Experimental validations on the real iQIYI video dataset have demonstrated the superior performance of our HCTRL method.

Acknowledgements. This work is supported in part by the National NSFC (Grants No. 62102053, 62272129, and 62372072), the Natural Science Foundation of Chongqing, China (Grant No. CSTB2022NSCQ-MSX1104), the General Program of Chongqing Science & Technology Commission (Grants No. CSTB2022TIAD-GPX0017, and CSTB2022TIAD-STX0006), the Science and Technology Plan Project of Chongqing Economic and Information Commission (Grant No. 2211R49R03), the Key Research Program of Chongqing Science & Technology Commission (Grant No. cstc2021jscx-dxwtBX0019), the Regional Innovation Cooperation Project of Sichuan Province (Grant No. 2023YFQ0028), and the Regional Science and Technology Innovation Cooperation Project of Chengdu City (Grant No. 2023-YF11-00023-HZ).

References

1. Lee, D., et al.: On the existence of a spectrum of policies that subsumes the least recently used (LRU) and least frequently used (LFU) policies. In: Proceedings of the International Conference on Measurement and Modeling of Computer Systems, pp. 134–143 (1999)
2. Bramson, M.: Instability of FIFO queueing networks. Ann. Appl. Probab. **4**(2), 414–431 (1994)
3. Yasir, M., uz Zaman, S.K., Maqsood, T., Rehman, F., Mustafa, S.: CoPUP: content popularity and user preferences aware content caching framework in mobile edge computing. Cluster Computing **26**(1), 267–281 (2023)
4. Srivastava, S., Singh, P.K.: Proof of optimality based on greedy algorithm for offline cache replacement algorithm. Int. J. Next-Gener. Comput. **13**(3) (2022)
5. Li, Y., Jiang, M., Zhang, Q., Qin, J.: Cache content placement optimization in non-orthogonal multiple access networks. IEEE Trans. Commun. **68**(7), 4580–4591 (2020)
6. Si Salem, T., Neglia, G., Ioannidis, S.: No-regret caching via online mirror descent. ACM Trans. Model. Perform. Eval. Comput. Syst. **8**(4), 1–32 (2023)
7. Mehrizi, S., Tsakmalis, A., Chatzinotas, S., Ottersten, B.: A feature-based bayesian method for content popularity prediction in edge-caching networks. In: IEEE Wireless Communications and Networking Conference, pp. 1–6 (2019)
8. Doan, K.N., Van Nguyen, T., Quek, T.Q.S., Shin, H.: Content-aware proactive caching for backhaul offloading in cellular network. IEEE Trans. Wirel. Commun. **17**(5), 3128–3140 (2018)

9. Qiu, S., Fan, Q., Li, X., Zhang, X., Min, G., Lyu, Y.: OA-Cache: oracle approximation based cache replacement at the network edge. IEEE Trans. Netw. Serv. Manage. (2023)
10. Zhong, C., Gursoy, M.C., Velipasalar, S.: A deep reinforcement learning-based framework for content caching. In: Annual Conference on Information Sciences and Systems, pp. 1–6 (2018)
11. Pingyang, W., Li, J., Shi, L., Ding, M., Cai, K., Yang, F.: Dynamic content update for wireless edge caching via deep reinforcement learning. IEEE Commun. Lett. **23**(10), 1773–1777 (2019)
12. Sadeghi, A., Sheikholeslami, F., Marques, A.G., Giannakis, G.B.: Reinforcement learning for adaptive caching with dynamic storage pricing. IEEE J. Sel. Areas Commun. **37**(10), 2267–2281 (2019)
13. Zhong, C., Gursoy, M.C., Velipasalar, S.: Deep reinforcement learning-based edge caching in wireless networks. IEEE Trans. Cogn. Commun. Networking **6**(1), 48–61 (2020)
14. Xiong, Z., Zhang, Y., Niyato, D., Deng, R., Wang, P., Wang, L.-C.: Deep reinforcement learning for mobile 5g and beyond: fundamentals, applications, and challenges. IEEE Veh. Technol. Mag. **14**(2), 44–52 (2019)
15. Vietri, G., et al.: Driving cache replacement with ML-based LeCaR. In: USENIX Workshop on Hot Topics in Storage and File Systems (2018)
16. Megiddo, N., Modha, D.S.: ARC: a self-tuning, low overhead replacement cache. In: 2nd USENIX Conference on File and Storage Technologies (2003)
17. Rodriguez, L.V., et al.: Learning cache replacement with CACHEUS. In: USENIX Conference on File and Storage Technologies, pp. 341–354 (2021)
18. Belady, L.A.: A study of replacement algorithms for a virtual-storage computer. IBM Syst. J. **5**(2), 78–101 (1966)

Multivariate Time Series Anomaly Detection Based on Device-Edge-Cloud Collaboration in Internet of Things

Yinkang Xu[1], Haowei Li[1], Zhiying Xiong[1], Qilin Fan[1(✉)], Xiuhua Li[1,2], and Cheng Zhang[3(✉)]

[1] School of Big Data and Software Engineering, Chongqing University, Chongqing 400044, China
xuyinkang@stu.cqu.edu.cn, {lihaowei,xiongzhiying,fanqilin}@cqu.edu.cn
[2] Haihe Laboratory of Information Technology Application Innovation, Tianjin 300072, China
[3] Chongqing Institute of Quality and Standardization, Chongqing 400023, China
zhch@cqis.cn

Abstract. The proliferation of the Internet of Things (IoT) has led to a significant surge in the quantity of multivariate time series (MTS) data. Effective detection of anomalies within MTS data is crucial for modern industrial applications. However, existing anomaly detection methods face challenges due to the lack of labeled data, imbalanced data distributions, and complex temporal and metric relationships inherent in MTS data. Moreover, in IoT network environments characterized by heterogeneous resource constraints, achieving high accuracy in anomaly detection while minimizing latency and energy consumption remains a daunting task. To address these challenges, in this paper, we propose a Collaborative Anomaly Detection (ColAD) approach to optimize the trade-off between latency, energy consumption, and accuracy. Specifically, we design a fusion of parallel graph attention networks with temporal convolutional attention networks to learn the correlations within MTS data across different metrics and temporal instances via attention mechanisms. Furthermore, a variational autoencoder model with dynamic depth-awareness mechanism is proposed to accommodate heterogeneous resources. Additionally, ColAD leverages a deep reinforcement learning based selection method to dynamically determine the optimal layer for processing MTS anomaly detection tasks. Experimental evaluations demonstrate the efficacy of ColAD in achieving superior performance in terms of anomaly detection accuracy, latency, and energy consumption.

Keywords: Multivariate Time Series · Anomaly Detection · Device-Edge-Cloud Collaboration · Task Offloading

Y. Xu and H. Li—Contributed equally to this work.

X. Li et al. (Eds.): 6GN 2024, LNICST 615, pp. 98–110, 2025.
https://doi.org/10.1007/978-3-031-93193-2_8

1 Introduction

Amidst the swift advancement of diverse sensors within the Internet of Things (IoT), substantial volumes of multivariate time series (MTS) data have been generated and collected across a multitude of applications during the systems running time. These applications span a spectrum including but not limited to environmental monitoring, manufacturing systems, and financial management. Anomaly detection for MTS identifies irregular patterns or events that deviate significantly from expected behavior, thus playing a pivotal role in facilitating proactive interventions, mitigating risks, and optimizing resource allocation.

Many traditional machine learning methods have been utilized to address this issue. Li *et al.* [1] devised a training method wherein a random forest classifier is employed to automatically determine suitable detector-parameter combinations and thresholds via labels. Nevertheless, the scarcity and difficulty in obtaining anomaly labels limit the applicability of this supervised learning method. Breunig *et al.* [2] proposed a distance-based method that operates on the premise that anomalous data points exhibit greater isolation compared to normal ones. However, its reliance on a global threshold poses challenges when dealing with datasets originating from disparate density regions. ECOD [3] calculates the empirical cumulative distribution to estimate the tail probabilities for each data dimension, and subsequently aggregates these probabilities to calculate an outlier score. Li *et al.* [4] employed a non-parametric technique by fitting empirical cumulative distribution functions to predict the tail probabilities of individual data points, thereby assessing their degree of "extremeness". However, these methods heavily rely on specific distribution assumptions. When MTS data deviates from these assumptions, their performance can decline.

Deep learning methods, due to their ability to capture complex feature relationships without the need for intricate feature engineering designs, and their capability for end-to-end training, have been widely employed in MTS anomaly detection in recent years. GDN [5] adopts an attention-based graph neural network to learn the dependence relationships between metrics. OmniAnomaly [6] learns a robust representation of normal patterns of MTS via stochastic variable connection and planar normalizing flow and determines anomalies through reconstruction probabilities. However, these methods primarily focus on the interplay among metrics, lacking in-depth exploration of temporal dependencies. Du *et al.* [7] proposed a generative adversarial network based method to filter possible anomalous samples with pseudo-labels before training the discriminator. Nonetheless, the additional component could be influenced by the anomaly pollution, resulting in increased errors in the pseudo-labeling. MAD-GAN [8] utilizes a long short-term memory (LSTM) to construct a generator and discriminator, simultaneously considering reconstruction error and discrimination error when calculating anomaly scores. However, this method fails to provide fast inference and high precision when encountering long time series.

Furthermore, considering the resource heterogeneity in IoT environment [9], it is challenging to meet the requirements of high accuracy, low latency, and energy efficiency for various MTS anomaly detection tasks solely by deploying an aforementioned model on the cloud, edge, or device layer. Therefore, a few network collaboration-based schemes have been proposed. HEC [10] constructs multiple deep neural network (DNN) models based on an autoencoder and LSTM with increasing complexity and associates each of them with hierarchical edge computing layers. Then an adaptive model selection scheme is designed. IOSARIMA [11] proposes a cloud-edge coordinated network traffic anomaly detection approach, deploying anomaly traffic alert models in edge regions and an anomaly traffic analysis model in the cloud server. ADRIoT [12] depends on the edge layer to execute anomaly detection, where the detector specific to a particular type of device is retrieved from the cloud. However, existing works have not fully exploited the potential of device-edge-cloud collaboration for MTS anomaly detection.

To this end, in this paper, we propose a **Col**laborative **A**nomaly **D**etection method (**ColAD**) for MTS in the device-edge-cloud network architecture. Our contributions are summarized as follows:

- We propose an integration of parallel graph attention networks (GATs) with temporal convolutional attention networks (TCANs), aimed at enhancing the modeling of correlations within MTS data across distinct metrics and temporal instances through an attention mechanism.
- We construct a multi-exit architecture based on variational autoencoder (VAE) to accommodate the hierarchical network layers. Besides, we leverage the deep reinforcement learning (DRL) based selection strategy to adaptively offload the anomaly detection tasks.
- Experimental results demonstrate that the proposed ColAD approach outperforms other candidate algorithms in terms of latency, energy consumption, and accuracy.

2 System Model and Problem Formulation

2.1 MTS Anomaly Detection Tasks Definition

There are continuous MTS anomaly detection tasks $\mathcal{A} = \{A_1, \ldots, A_t, \ldots\}$ generated by IoT devices within the coverage area. Let s_t and c_t denote the data size and required computation period of A_t, respectively. Each task A_t takes a sliding window of MTS $w_t \in \mathbb{R}^{L \times D}$, where L is the time window length and D is the number of collected and monitored metrics. If there is an anomalous event, the task is to produce a final output $y_t \in \{0, 1\}$, where $y_t = 1$ indicates the detection of an anomaly, and 0 otherwise.

2.2 Network Architecture

We leverage a hierarchical device-edge-cloud network architecture [13] for MTS anomaly detection task, which is divided into three layers:

- Device Layer: It is the bottom layer of the network that consists of massive amounts of IoT devices. The primary responsibilities of the device layer encompass the collection, preprocessing, and initial anomaly detection of raw data. However, the device layer may face challenges due to the computational intensity associated with anomaly detection tasks under limited computing/storage capability.
- Edge Layer: It is the intermediate layer between the device layer and cloud layer which consists of edge servers (deployed at base stations, access points, or edge routers). To alleviate the load on the cloud and further reduce the end-to-end latency, edge servers can be leveraged to carry out anomaly detection tasks of intermediate complexity that necessitate moderate computing/storage resources.
- Cloud Layer: It is the top layer of the network. The cloud servers are remote from devices, which all the edge servers are connected to by extremely large capacity wired networks. Compared with edge servers, cloud servers have much more powerful computing/storage resources to train complex models for undertaking computation-intensive tasks.

2.3 Task Offloading

In terms of task allocation, we set $b_t = [\lambda_d, \lambda_e]$ as the offloading variable vector of the three-layer network structure. The binary variable $\lambda_d = 0$ indicates that task A_t is executed locally, and $\lambda_d = 1$ indicates that the task is offloaded to the edge server. The binary variable $\lambda_e = 0$ indicates that the task is executed on the edge server, while $\lambda_e = 1$ indicates that the task is offloaded to the cloud server.

If $b_t = [0, 0]$, the task is directly executed by the device layer. If $b_t = [1, 0]$, the task is compressed by the device layer and transmitted to the edge layer for processing to obtain the detection result. Let η_t^{de} represent the data compression ratio of the device layer encoder, then the compressed data size s_t^{de} of task A_t received by the edge layer is $\eta_t^{de} s_t$.

If $b_t = [1, 1]$, the task is transmitted to the cloud layer for processing, and the cloud layer performs encoding and decoding reconstruction to obtain the detection result. Similarly, Let η_t^{ec} represent the data compression ratio of the edge layer encoder, then the compressed data size s_t^{ec} is $\eta_t^{ec} s_t^{de}$.

2.4 Delay Calculation

The delay t_t^{all} comes from the processing delay t_t^{proc} and transmission delay t_t^{trans}. If $b_t = [0, 0]$, the procedure entails the initial compression of task A_t by the encoder, followed by a decision made by the offloading model regarding its local execution. Subsequently, the task A_t undergoes decoding. In this case, $t_t^{trans} = 0$. The t_t^{all} is equivalent to the processing delay which contains three parts: encoding delay, decision delay, and decoding delay:

$$t_t^{all} = t_t^{proc} = \frac{(1 + 2\eta_t^{de})c_t}{C_d + \varepsilon_0}, \tag{1}$$

where C_d represents the computing capacity of the device layer. $\eta_t^{de} c_t$ represents the computing cycles required after the data is encoded by the device layer encoder. The offloading decision and the input of the decoder are the same, so we let the decision and decoding cycles be $2\eta_t^{de} c_t$. In addition, $\varepsilon_0 = 10^{-9}$ is a small value parameter introduced to avoid zero denominators. The $D_{\text{in-device}}$ and $D_{\text{out-device}}$ represent the input and output dimensions of the data in the device layer, respectively. Therefore, the compression ratio η_t^{de} is $\frac{D_{\text{in-device}}}{D_{\text{out-device}}}$.

If $b_t = [1, 0]$, the task is transmitted to the edge layer, the processing delay includes the processing delay of the device and the edge layers. The processing delay of the device layer comprises encoding and decoding delays, while the processing delay of the edge layer comprises encoding, decision, and decoding delays. Consequently, the t_t^{proc} is expressed as follows:

$$t_t^{proc} = \frac{(1 + \eta_t^{de}) c_t}{C_d + \varepsilon_0} + \frac{(1 + 2\eta_t^{ec}) \eta_t^{de} c_t}{C_e}, \tag{2}$$

where C_e represents the computing capacity of the edge server. Similarly, $D_{\text{out-device}}$ and $D_{\text{out-edge}}$ denote the dimensions of output by the encoder of the device layer and edge layer respectively. The compression ratio η_t^{de} can be calculated by $\frac{D_{\text{out-device}}}{D_{\text{out-edge}}}$.

When the device layer proposes a transmission request, the task is sent to the edge server via a wireless link. The transmission delay only includes the transmission delay from the device layer to the edge layer:

$$t_t^{trans} = \frac{\eta_t^{de} s_t}{R_t^{de}}, \tag{3}$$

where R_t^{de} denotes the transmission rate from the device layer to the edge layer. We calculate the $R_t^{de} = B_{de} log_2(1 + \frac{p_{de}|h_{de}|^2}{\sigma_{dc}^2})$, and B_{de}, p_{de}, h_{de}, and σ_{de}^2 respectively represent the transmission bandwidth, transmission power, channel gain, and noise power between the device layer and the edge layer.

If $b_t = [1, 1]$, the task is transmitted to the cloud layer, the processing delay includes not only the processing delay of the device layer and the edge layer but also the processing delay of the cloud layer. The processing delay of the cloud layer includes encoding delay and decoding delay. The formula of t_t^{proc} is given by:

$$t_t^{proc} = \frac{(1 + \eta_t^{de}) c_t}{C_d + \varepsilon_0} + \frac{(1 + \eta_t^{ec}) \eta_t^{de} c_t}{C_e} + \frac{(1 + \eta_t^{cc}) \eta_t^{de} \eta_t^{ec} c_t}{C_c}, \tag{4}$$

where C_c represents the computing capacity of the cloud server.

The transmission delay includes delays from both the device layer to the edge layer and the edge layer to the cloud layer. The edge server connects to the cloud server via high-capacity wired optical cables with a transmission rate of R_t^{ec}. The total transmission delay t_t^{trans}, is calculated as

$$t_t^{trans} = \frac{\eta_t^{de} s_t}{R_t^{dc}} + \frac{\eta_t^{de} \eta_t^{ec} s_t}{R_t^{ec}}. \tag{5}$$

In summary, the total delay for MTS anomaly detection in the device-edge-cloud network architecture is:

$$t_t^{all} = \begin{cases} \frac{(1+2\eta_t^{de})c_t}{C_d+\varepsilon_0}, & \text{if } b_t = [0,0]; \\ \frac{(1+\eta_t^{de})c_t}{C_d+\varepsilon_0} + \frac{(1+2\eta_t^{ec})\eta_t^{de}c_t}{C_e} + \frac{\eta_t^{de}s_t}{R_t^{de}}, & \text{if } b_t = [1,0]; \\ \frac{(1+\eta_t^{de})c_t}{C_d+\varepsilon_0} + \frac{(1+\eta_t^{ec})\eta_t^{de}c_t}{C_e} + & \\ \frac{(1+\eta_t^{cc})\eta_t^{de}\eta_t^{ec}c_t}{C_c} + \frac{\eta_t^{de}s_t}{R_t^{de}} + \frac{\eta_t^{de}\eta_t^{ec}s_t}{R_t^{ec}}. & \text{if } b_t = [1,1]. \end{cases} \tag{6}$$

2.5 Energy Consumption Calculation

The total energy consumption includes both computational and transmission energy. Therefore, the energy consumption of a task is calculated as follows:

$$p_t^{all} = \begin{cases} \frac{(1+2\eta_t^{de})c_t}{C_d+\epsilon_0}P_d, & b_t = [0,0]; \\ \frac{(1+\eta_t^{de})c_t}{C_d+\epsilon_0}P_d + \frac{(1+2\eta_t^{ec})\eta_t^{de}c_t}{C_e}P_e & \\ +\frac{\eta_t^{de}s_t}{R_t^{de}}P_{de}, & b_t = [1,0]; \\ \frac{(1+\eta_t^{de})c_t}{C_d+\epsilon_0}P_d + \frac{(1+\eta_t^{ec})\eta_t^{de}c_t}{C_e}P_e & \\ +\frac{(1+\eta_t^{cc})\eta_t^{de}\eta_t^{ec}c_t}{C_c})P_c & \\ +\frac{\eta_t^{de}s_t}{R_t^{de}}P_{de} + \frac{\eta_t^{de}\eta_t^{ec}s_t}{R_t^{ec}}P_{ec}, & b_t = [1,1], \end{cases} \tag{7}$$

where P_d, P_e, P_c, P_{de}, P_{ec} represent the average power consumption of the device layer, edge layer, cloud layer, device-to-edge, and edge-to-cloud, respectively.

2.6 Problem Formulation

In anomaly detection tasks, latency, energy consumption, and accuracy are all crucial. We use the F1 score to measure a model's accuracy. Here we design Latency-Energy-F1 (LEF) as the performance metric, aiming to minimize latency and energy consumption while maximizing the F1 score:

$$\text{LEF} = \alpha_1 \frac{t_{max} - t^{all}}{t_{max} - t_{min}} + \alpha_2 \frac{p_{max} - p^{all}}{p_{max} - p_{min}} + \alpha_3 \text{F1}, \tag{8}$$

where t^{all} and p^{all} denote the total latency and total energy consumption. t_{max} and p_{max} represent the maximum values of total latency and total energy consumption, respectively. t_{min} and p_{min} denote the minimum values of total latency and total energy consumption. α_1, α_2, and α_3 are weighting factors.

3 The Design of ColAD Algorithm

The pipeline of the proposed ColAD for anomaly detection is shown in Fig. 1. We design a DNN based on multi-exit architecture of VAE model. The VAE model is split into two parts (i.e., the encoder and the decoder), where each decoder

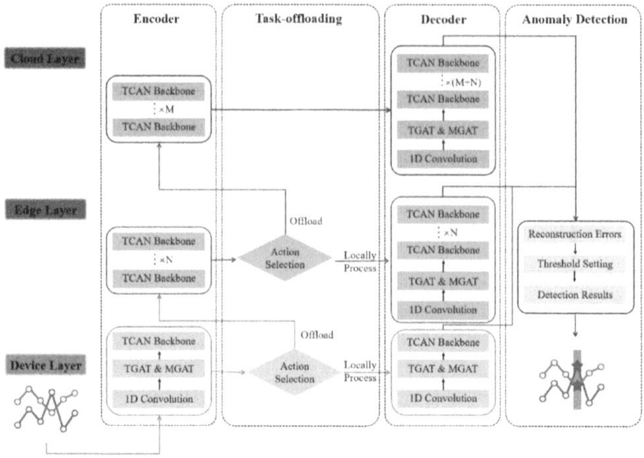

Fig. 1. The pipeline of ColAD algorithm. M and N denote the number of TCAN backbones in the edge layer and cloud layer, respectively.

can be considered as an exit. By combining parallel GATs and TCANs, ColAD adapts to variations across different temporal scales and effectively captures the interdependencies of MTS across both metric and temporal dimensions. Furthermore, ColAD utilizes a DRL-based method to make adaptive decisions on whether the task is offloaded to the next layer (except for the last layer) or output by the current exit and locally processed. Next, we elaborate on the main components.

3.1 Parallel Graph Attention Networks

We first utilize a one-dimensional convolution with a kernel size of k to extract advanced feature c_t from w_t, where $c_t = [c_t^1, \ldots, c_t^D]$. Let $r_t = [r_t^1, \ldots, r_t^L]$ denote the transpose of c_t.

Then, c_t or r_t is conceptualized as a comprehensive graph, where nodes symbolize features and edges represent the relationships among features. We leverage parallel GATs, denoted as metric graph attention network (MGAT) $G_t = \{V_t, E_t\}$ and temporal graph attention network (TGAT) $\hat{G}_t = \{\hat{V}_t, \hat{E}_t\}$ to discern metric-oriented and time-oriented correlations, respectively. Specifically, $V_t = c_t$, $E_t \subseteq V_t \times V_t$; $\hat{V}_t = r_t$, $\hat{E}_t \subseteq V_t \times V_t$. We compute the attention coefficient α_{ij} between node c_t^i and node c_t^j and the attention coefficient $\hat{\alpha}_{ij}$ between node r_t^i and node r_t^j as follows:

$$\begin{aligned} \alpha_{ij} &= \text{Softmax}(\text{LeakyReLU}(\boldsymbol{a}_1^\top \cdot (W_1 c_t^i \parallel W_1 c_t^j))), \\ \hat{\alpha}_{ij} &= \text{Softmax}(\text{LeakyReLU}(\boldsymbol{a}_2^\top \cdot (W_2 r_t^i \parallel W_2 r_t^j))), \end{aligned} \tag{9}$$

where $\boldsymbol{a}_1^T$ and $\boldsymbol{a}_2^T$ are parameter vectors, W_1 and W_2 are weight matrices, and $\parallel$ denotes the connection operation. So, the dependency relationship f_t of metrics and the dependency relationship $\hat{f}_t$ of time are given by:

$$f_t = \mathrm{MGAT}(G_t) = \sum\nolimits_{i=1}^{D} (\mathrm{Sigmoid}(\sum\nolimits_{j=1}^{D} \alpha_{ij} c_t^i)),$$
$$\hat{f}_t = \mathrm{TGAT}(\hat{G}_t) = \sum\nolimits_{i=1}^{L} (\mathrm{Sigmoid}(\sum\nolimits_{j=1}^{L} \hat{\alpha}_{ij} r_t^i)). \quad (10)$$

Finally, we obtain the aggregated feature h_t by aggregating c_t, f_t and $\hat{f}_t$.

3.2 Temporal Convolutional Attention Network

Temporal convolutional network (TCN) [14] is well-suited for modeling time series due to its computational efficiency and ability to capture long-range dependencies. To flexibly accommodate variations across different temporal scales and effectively characterize intricate dependency patterns within MTS, we enhance each residual module of the TCN by adding a self-attention mechanism. So, each residual module is constructed with dilated convolutional layers, weight normalization layers, ReLU activation functions, dropout layers, and self-attention layers. The output of the residual module is given by:

$$o_t = \mathrm{Sigmoid}(\mathrm{Conv1D}(h_t) + F(h_t)), \quad (11)$$

where $F(\cdot)$ is the residual function. Specifically, the scaled-dot product attention of self-attention layer is defined as:

$$a_t = \mathrm{Softmax}(\frac{q_t {k_t}^{\mathsf{T}}}{\sqrt{d_k}}) v_t, \quad (12)$$

where q_t, k_t and v_t represent query, key and value matrices transformed from o_t with parameters, respectively. d_k is the key dimension, serving as a scaling factor.

3.3 VAE Model with Dynamic Depth-Awareness Mechanism

The VAE is a deep Bayesian model for learning the data distribution. It learns a joint distribution $p_\psi(w_t, z_t)$ which is decomposed as $p_\psi(w_t, z_t) = p_\psi(z_t) p_\psi(w_t|z_t)$ with a prior distribution over latent space $p_\psi(z_t)$, and a decoder $p_\psi(w_t|z_t)$. The encoder $q_\phi(z_t|w_t)$ approximates the true but intractable posterior $p_\psi(z_t|w_t)$. VAE is trained by maximizing the evidence lower bound considering the reconstruction and generalization capabilities. The VAE model in ColAD exhibits a multi-exit architecture featuring a dynamic depth-awareness mechanism, which is designed to accommodate the heterogeneous resources encompassing device, edge, and cloud layers, thereby facilitating collaborative anomaly detection. The detailed modules and connections are depicted in Fig. 1.

3.4 Task Offloading Strategy Based on DQN

DRL shows its potential in adjusting the strategy via the interaction between the agent and environment for maximizing a predetermined reward value based on the system's state [15]. Therefore, we utilize deep Q-network (DQN) as the model selection scheme to balance the latency, energy consumption, and accuracy for MTS anomaly detection under the hierarchical heterogeneous resources. ColAD consists of two DQN models deployed at the device and edge layer, respectively.

For task A_t, the agent observes the state s_t and selects a discrete action a_t. After executing this action a_t, the agent receives a reward r_t and observes the next state s_{t+1}. Specifically, the current state s_t is defined as the hidden representation from the encoder. If the current action equals one, the current state is transferred to the upper layer without decoder processing; otherwise, the output is taken from the exit of the current layer (i.e., locally processed by the decoder). We adopt an ϵ-greedy algorithm for action selection. The agent will randomly select an action from the action space with a probability of $P = \epsilon$, and with a probability of $P = 1 - \epsilon$, it will choose the action that maximizes the action-state value function $Q(s_t, a_t)$. We use the difference between raw data and reconstructed data (i.e., reconstruction error) as a criterion for evaluating the model's accuracy. Therefore, combined with latency and energy consumption, the reward is given by:

$$r_t = \alpha_1 \cdot \frac{1}{t_t^{\text{all}}} \cdot +\alpha_2 \cdot \frac{1}{p_t^{\text{all}}} + \alpha_3 \cdot \left\| w_t - w_t^{\text{rec}} \right\|, \tag{13}$$

where w_t^{rec} is the reconstruction value.

The optimal state-action value function $Q(s_t, a_t)$ is defined as $Q(s_t, a_t; \theta)$, where θ represents the parameters of the DNN, and the update rule is as follows:

$$\begin{gathered} \theta_{t+1} = \theta_t + \\ \alpha \left[r_t + \gamma_{a_t'}^{max} Q(s_t', a_t', \theta^-) - Q(s_t, a_t; \theta) \right] \nabla Q(s_t, a_t; \theta) \end{gathered} \tag{14}$$

The loss function $L(\theta)$ measures the difference between the network's predicted value and the real value provided by the target network to guide parameter updates:

$$L(\theta) = E\left[\left(r_t + \gamma_{a_t'}^{max} Q(s_t', a_t', \theta^-) - Q(s_t, a_t; \theta) \right)^2 \right]. \tag{15}$$

The update of the action-state value function follows the Bellman expectation equation:

$$\mathcal{Q}_{t+1}(s_t, a_t) = E_{s'}[r_t + \gamma_{a_t'}^{max} Q_t(s_t', a_t')]. \tag{16}$$

4 Evaluation

4.1 Experimental Setup

Dataset. We conduct experiments using the MBD dataset, which originates from an environment that operates a big data batch processing system. It contains one master and four slaves, with each node monitoring 26 metrics. There are 8200 timestamps with an anomaly ratio of 7.0556%. The dataset is fully labeled.

Implementation. The computing capacities of each device are randomly assigned within a range of 1 to 4 GHz, while each edge server and cloud are allocated 30 GHz and 100 GHz, respectively. The transmission bandwidth is set at 5 MHz, the transmission power at 20 dBm, and the noise power at -102 dBm. The large-scale fading in the channel model mainly depends on the path loss and log-normal shadowing between IoT devices and edge servers, while the small-scale fading is Rayleigh fading with unit variance. The model is implemented using PyTorch. The sliding window, batch size, slide stride, and greedy factor are set to 25, 20, 7, and 0.95, respectively. The numbers of TCAN backbones in the edge layer and cloud layer are set to 4 and 8, respectively.

Baselines. We compare the effectiveness of the ColAD algorithm with the following approaches: 1) **HEC** [10]; 2) **IOSARIMA** [11]; 3) **ADRIoT** [12].

4.2 Results and Analysis

Table 1. Performance Comparison on MBD Dataset.

	Recall	Precision	F1 score	LEF
HEC	0.9309	0.8840	0.9392	0.8695
IOSARIMA	0.8643	0.9434	0.8713	0.8544
ADRIoT	0.7159	0.7702	0.8339	0.8376
ColAD	0.9160	0.9752	**0.9567**	**0.8868**

Performance Comparison. We use precision, recall, F1 score, and LEF as performance metrics. Our proposed method, as shown in Table 1, exhibits excellent ability, achieving the highest LEF metric.

Unlike ColAD which segments the complex DNN model into several parts for hierarchical layers with distinct capabilities, HEC constructs three independent DNN models of increasing complexities in device, edge, and cloud layers. Although HEC achieves a high F1 score due to its robust model, the high latency and energy consumption lead to a lower LEF score. IOSARIMA initially filters data using an anomaly detection model at the edge layer and then deploys an anomaly analysis model at the cloud layer, reducing data transmission between the edge and cloud layers to some extent. However, the complex collaboration between layers results in an overall unsatisfactory LEF metric. ADRIoT abstains from deploying models at the device layer and instead depends on the edge layer to retrieve detectors from the cloud for local execution. This approach entails substantial communication requirements and limits the utilization of the cloud's powerful computational capabilities for handling intricate tasks, consequently yielding comparatively subpar performance in both F1 score and LEF. In summary, our proposed ColAD exhibits notable advantages in terms of generalization capability and efficient resource utilization, thereby attaining exceptional comprehensive performance metrics.

Ablation Study. To validate the effectiveness of the components proposed in ColAD, we test the following variants of ColAD: 1) **w/o DQN**, representing the model that employs a random strategy for offloading as an alternative instead of DQN; 2) **w/o Successive**, representing the model that data is transmitted directly from the device layer to the cloud layer, circumventing the intermediate edge layer. 3) **w/o Partition**, representing the model discards a multi-exit architecture, where each layer constitutes a complete VAE model.

Table 2. Ablation Experiments on Recall, Precision, F1 Score and LEF metric of ColAD and its variants.

	Recall	Precision	F1 score	LEF
w/o DQN	0.8514	0.9850	0.9133	0.8631
w/o Successive	0.8656	0.9738	0.9265	0.8687
w/o Partition	0.9341	0.9757	0.9544	0.8573
ColAD	0.9160	0.9752	**0.9567**	**0.8868**

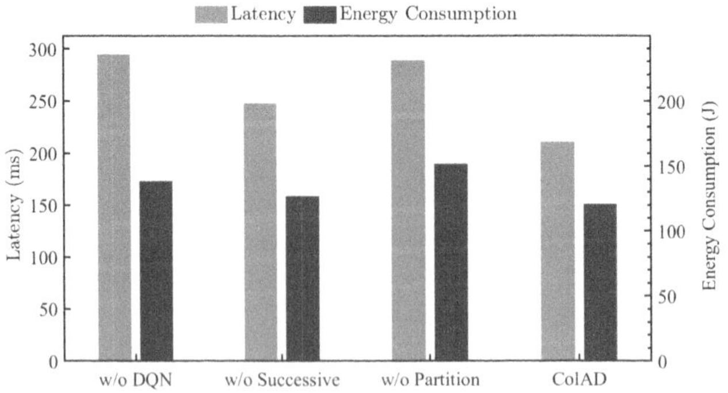

Fig. 2. Comparison of latency and energy consumption metrics of ColAD and its variants.

Table 2 and Fig. 2 illustrate the comparison between ColAD and its variants. The w/o DQN suffers from performance degradation compared to the proposed ColAD. The results show that DQN can provide a better balance among latency, energy consumption, and accuracy to choose an intelligent offloading strategy, verifying the necessity of the decision-making module. The w/o Successive bypasses the edge layer, which leads to anomaly detection tasks that should have been offloaded to the edge layer being offloaded to the cloud layer instead to some extent. The decreased LEF performance shows that the successive hierarchical device-edge-cloud offloading order is essential. When removing the partition module, in which the encoder and decoder are regarded as an indivisible component without a multi-exit mechanism, uncompressed data, and redundant computations could lead to increased latency and energy consumption, consequently reducing LEF. This highlights the potential of DNN model partition with a multi-exit mechanism to facilitate deep collaboration among heterogeneous resources.

5 Conclusions

In this paper, we have investigated the issue of MTS anomaly detection based on device-edge-cloud collaboration in the IoT scenario. We have presented an integration of parallel GATs with TCANs to enhance the modeling of correlations within MTS data across different metrics and temporal instances through attention mechanism. We also have proposed a multi-exit architecture based on VAE to accommodate hierarchical network layers. Additionally, we have utilized a DQN-based based selection strategy to adaptively offload anomaly detection tasks. The experimental results show that ColAD outperforms other compared approaches.

Acknowledgements. This work is supported in part by the National NSFC (Grants No. 62102053, 62272129, and 62372072), the Natural Science Foundation of Chongqing, China (Grant No. CSTB2022NSCQ-MSX1104), the General Program of Chongqing Science & Technology Commission (Grants No. CSTB2022TIAD-GPX0017, and CSTB2022TIAD-STX0006), the Science and Technology Plan Project of Chongqing Economic and Information Commission (Grant No. 2211R49R03), the Key Research Program of Chongqing Science & Technology Commission (Grant No. cstc2021jscx-dxwtBX0019), the Regional Innovation Cooperation Project of Sichuan Province (Grant No. 2023YFQ0028), and the Regional Science and Technology Innovation Cooperation Project of Chengdu City (Grant No. 2023-YF11-00023-HZ).

References

1. Liu, D., et al.: Opprentice: towards practical and automatic anomaly detection through machine learning. In: Proc. IMC, pp. 211–224 (2015)
2. Breunig, M.M., Kriegel, H.P., Ng, R.T., Sander, J.: LOF: identifying density-based local outliers. In: Proc. SIGMOD, pp. 93–104 (2000)

3. Li, Z., Zhao, Y., Hu, X., Botta, N., Ionescu, C., Chen, G.H.: ECOD: unsupervised outlier detection using empirical cumulative distribution functions. IEEE Trans. Knowl. Data Eng. **35**(12), 12181–12193 (2023)
4. Li, Z., Zhao, Y., Botta, N., Ionescu, C., Hu, X.: COPOD: copula-based outlier detection. In: Proc. ICDM, pp. 1118–1123 (2020)
5. Deng, A., Hooi, B.: Graph neural network-based anomaly detection in multivariate time series. In: Proc. AAAI, pp. 4027–4035 (2021)
6. Su, Y., Zhao, Y., Niu, C., Liu, R., Sun, W., Pei, D.: Robust anomaly detection for multivariate time series through stochastic recurrent neural network. In: Proc. KDD, pp. 2828–2837 (2019)
7. Du, B., Sun, X., Ye, J., Cheng, K., Wang, J., Sun, L.: GAN-based anomaly detection for multivariate time series using polluted training set. IEEE Trans. Knowl. Data Eng. **35**(12), 12208–12219 (2023)
8. Li, D., Chen, D., Jin, B., Shi, L.: MAD-GAN: multivariate anomaly detection for time series data with generative adversarial networks. In: Proc. ICANN, pp. 703–716 (2019)
9. Li, P., et al.: Filling the missing: exploring generative AI for enhanced federated learning over heterogeneous mobile edge devices. IEEE Trans. Mob. Comput. (2024). https://doi.org/10.1109/TMC.2024.3371772
10. Ngo, V., Luo, T., Quek, T.Q.: Adaptive anomaly detection for internet of things in hierarchical edge computing: a contextual-bandit approach. ACM Trans. Internet Things **3(4)**(1), 1–23 (2021)
11. Yang, T., Hao, W., Yang, Q., Wang, W.: Cloud-edge coordinated traffic anomaly detection for industrial cyber-physical systems. Expert Syst. Appl. **230**(15), 120668 (2023)
12. Li, R., Li, Q., Zhou, J., Jiang, Y.: ADRIoT: An edge-assisted anomaly detection framework against IoT-based network attacks. IEEE Internet Things J. **9**(13), 10576–10587 (2022)
13. Li, X., Xu, Z., Fang, F., Fan, Q., Wang, X., Leung, V.C.: Task offloading for deep learning empowered automatic speech analysis in mobile edge-cloud computing networks. IEEE Trans. Cloud Comput. **11**(2), 1985–1998 (2023)
14. Bai, S., Kolter, J.Z., Koltun, V.: An empirical evaluation of generic convolutional and recurrent networks for sequence modeling. arXiv preprint arXiv:1803.01271 (2018)
15. Dai, Y., Zhang, K., Maharjan, S., Zhang, Y.: Deep reinforcement learning for stochastic computation offloading in digital twin networks. IEEE Trans. Industr. Inf. **17**(7), 4968–4977 (2021)

Intelligent Systems and Simulation

UAV Topology Control Method Based on Fuzzy C-Clustering Algorithm

Jiajie Guo[1], Jian He[1], Dawei Chen[2], and Shuo Shi[1](✉)

[1] Harbin Institute of Technology, Harbin 150001, China
crcss@hit.edu.cn
[2] Chinese Aeronautical Establishment, Beijing 100012, China

Abstract. As unmanned aerial vehicle(UAV) technology has evolved in recent years, the need for mission execution has evolved towards multi-UAV collaboration. The scale of UAV clusters has become larger and larger, then the network topology control method for UAV clusters is particularly important. In this paper, on the basis of researching a large amount of literature, an improved clustering algorithm based on the fuzzy C-clustering algorithm is proposed to be applied to the topology control of UAV networks, and relevant mathematical modelling and algorithmic simulation analyses are carried out. The analysis results show that the method used in this paper has better performance than the widely used Low Energy Adaptive Clustering Hierarchy(LEACH) algorithm. The use of this algorithm can show better topology control results when working together in UAV clusters.

Keywords: Topology control · Fuzzy C-clustering algorithm · LEACH algorithm

1 Introduction

UAV technology has been developing rapidly in recent years, and intelligence and informatization have become its remarkable development characteristics. Whether in the civil or military field, UAV technology can be applied in a variety of different scenarios. Such as search and rescue, remote communication, area surveillance [1], mobile edge computing and so on [2].

As UAV technology has evolved in recent years, the need for mission execution has evolved towards multi-UAV collaboration. The scale of UAV clusters has become larger and larger, so the network topology control method of UAV clusters is especially important. How to design an efficient topology control algorithm to better meet the specifications of UAV clusters in the process of executing actual missions is an important problem that needs to be solved urgently. Aiming at this problem, scholars at home and abroad have carried out a lot of work.

So far, research on topology control of self-organizing networks has been conducted for more than fifty years, and in the process, a wide variety of algorithmic categories have been derived. Currently, the mainstream differentiation directions are divided into three

X. Li et al. (Eds.): 6GN 2024, LNICST 615, pp. 113–124, 2025.
https://doi.org/10.1007/978-3-031-93193-2_9

main categories: connectivity and energy criterion-based algorithms, traffic equalization criterion-based algorithms, and stability criterion-based algorithms.

Firstly, the primary metrics of connectivity and energy criterion based algorithms are the degree of interconnectivity between nodes and minimizing the number of communication links established by the nodes so as to optimize the efficiency of their energy utilization. Literature [3] suggests that to preserve the connectivity of the network, the shortest communication distance between nodes needs to be equal to the longest edge in the minimum spanning tree and based on this the shortest inter-node communication distance is approximated using the longest Long Short-Term Memory (LMST) edge. Literature [4] states that in topology control networks, the use of two-hop or three-hop route discovery depending on the degree of nodes can reduce the number of routing packets and improve the operational efficiency of the network, and based on this, a locally adaptive self-configuring LMST topology control method is analyzed.

Traffic equalization criterion type algorithms consider the throughput of the wireless network during multichannel communication. Literature [5] proposes the concept of co-channel interference and seeks channel allocation for a given network such that the induced network topology minimizes the interference among all K-connected topologies, proposes a polynomial time optimal algorithm and maximum bottleneck capacity path routing heuristic that improves the system performance by an average of 57% in terms of connection blocking rate. Literature [6] proposes a power-aware cluster selection and reorganization method, which gives a reference solution for the optimization problem of network dynamic topology.

The stability criterion ensures smooth communication between nodes and excludes the phenomenon of inter-node failures. Literature [7] proposes a method to determine the location of relay nodes based on a virtual mesh for optimal deployment computation with full-area coverage for the phenomenon of faulty links and nodes, and proves that the method has good performance. Literature [8] proposes a topology control method based on positional motion and Deep Reinforcement Learning (DRL), which provides a good reference in training and deploying DRL algorithms on UAVs with limited resources and hardware constraints to ensure network connectivity and fault tolerance.

UAV network structure is complex, and the research of related network control technology has been on the rise. However, with the advancement of intelligence, the performance requirements of UAVs will become higher and higher; coupled with the increasingly large scale of UAV clusters and more and more scenarios of multiple UAVs working together, more efficient network topology control methods have become an important demand.

2 System Model

UAVs, as nodes in a network, have and their flexible mobility. To study the topology of UAV network, the movement of UAVs must be modeled first. The more widely used UAV cluster movement models are divided into two kinds: one is the single movement model, there is no mutual influence between individuals, and the speed and trajectory of their movement depend entirely on themselves; the other is the group movement model, which moves in a cluster formed by multiple nodes, and the nodes within the cluster are connected to each other and follow specific movement rules.

In this paper, considering the generality of the model, we choose the combination of random wandering model and group pursuit model to simulate the mission execution scenario under multi-UAV collaboration.

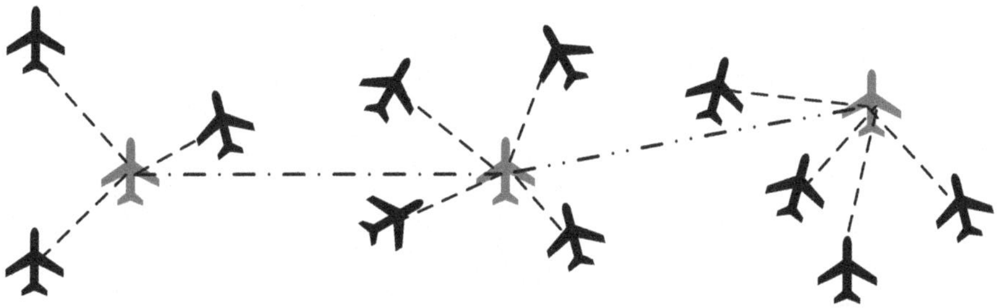

Fig. 1. Multiple UAVs working together

As shown in the Fig. 1, under the multi-UAV cluster cooperative work, the movement model of the pursued node is defined to be the same as the random wandering model, while the pursuing node moves at a certain speed according to the pursuing model.

2.1 Modeling of Random Walks

Firstly, in this paper, the initial positions of the nodes are randomly and uniformly preset in the boundary range, and the total number of rounds of motion is calculated by the running time and step time. In this way, the velocity magnitude and direction of all rounds of a single node are randomized. Set the node to move in its velocity magnitude range $[V_{min}, V_{max}]$, direction range $(0, 2\pi)$ uniformly and randomly selected, the movement step is Δt, the position of the node at the moment t can be expressed as:

$$x_t = x_{t-\Delta t} + v_{t-\Delta t} * cos\theta_t * \Delta t \quad (1)$$

$$y_t = y_{t-\Delta t} + v_{t-\Delta t} * sin\theta_t * \Delta t \quad (2)$$

where θ_t is the node motion velocity angle to the x-axis. Each node repeats the above process independently, and when the node collides with the boundary of the region, it reverses the velocity of the horizontal or vertical axis of the velocity, or takes the reverse of both, thus ensuring that the node does not exceed the established range of motion. The random wandering model simulation diagram is shown in Fig. 2, which shows that the nodes are doing irregular motion and will not go beyond the established range.

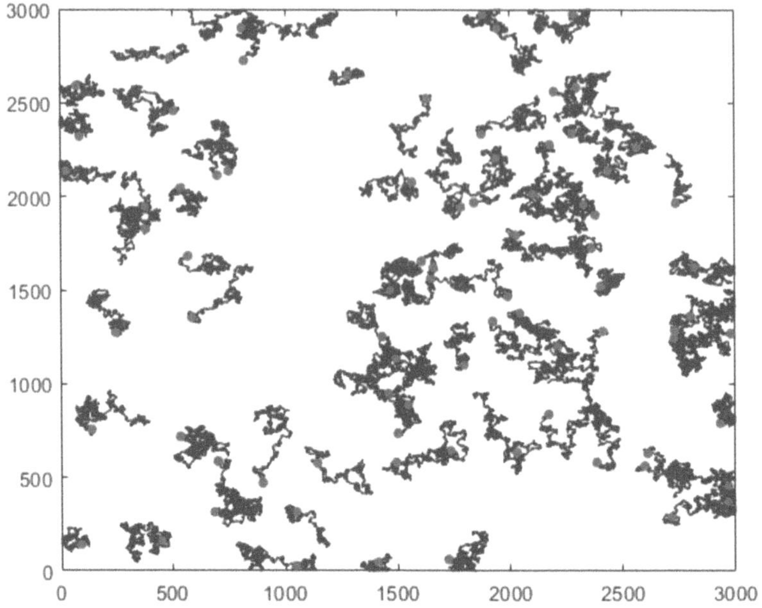

Fig. 2. Modeling of random walks

2.2 Modeling of Group Catching-Up

In this paper, the pursued nodes are set to move randomly, while the pursuing nodes move in a group pursuit model, and the speed magnitude of the pursuing nodes is higher than that of the pursued nodes, which mainly simulates the scenarios of multiple UAVs pursuing enemy aircrafts, and UAV encircling performances.

For the pursued node, its motion model has already been established and will not be repeated. For the chasing node, it is required that its velocity magnitude is still selected uniformly and randomly within $[V_{min}, V_{max}]$, and because the chasing node will not make a foldback, the change of its velocity direction should be satisfied:

$$\theta_{t+\Delta t} = \theta_t + a\Delta\theta \tag{3}$$

In which:

$$\Delta\theta\epsilon[-\pi, \pi], a\epsilon\left(0, \frac{1}{2}\right) \tag{4}$$

The iteration process $\Delta\theta$ obeys a uniform distribution. Each subsequent movement satisfies:

$$x_t = x_{t-\Delta t} + v_c * cos\theta_t * \Delta t \tag{5}$$

$$y_t = y_{t-\Delta t} + v_c * sin\theta_t * \Delta t \tag{6}$$

In which:

$$v_c \in [v_{min}, v_{max}] \tag{7}$$

The simulation results of the group catch-up model are shown in the Fig. 3:

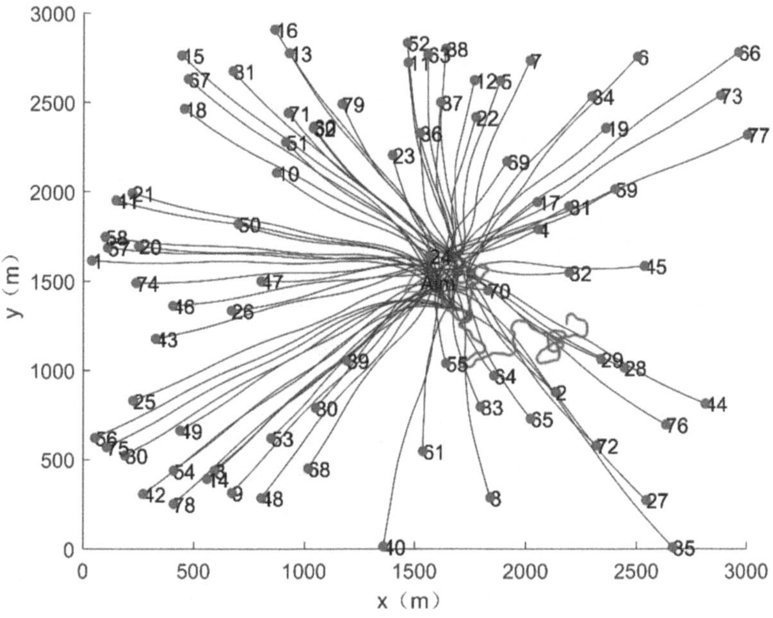

Fig. 3. Modeling of group catching-up

2.3 Design of Evaluation Indicators

(1) Load Balance

Ideally, nodes should be evenly distributed under these cluster heads, however, in practice, there are often more nodes in some clusters and fewer nodes in others, resulting in uneven distribution of communication resources. Therefore, this paper proposes the metric of load balancing degree to measure whether the communication resources are utilized efficiently or not:

$$LBM = \frac{m}{\sum_{i=0}^{m} \left(num\left(x_{j,j\in C_i}\right) - \overline{num} \right)^2} \tag{8}$$

where $\overline{num}$ denotes the average number of nodes under each sub-cluster, $num\left(x_{j,j\in C_i}\right)$ is the number of nodes under each sub-cluster, and m denotes the number of sub-clusters. The equation shows that the larger the load balancing degree is, the better the network communication resource allocation is about and the more reasonable the topology is.

(2) Average Interference Degree

As shown in Fig. 4, two nodes A and B are communicating and a circle is made with their communication radius, which represents the communication range of that node. If there are other nodes within the communication range of a node, these nodes are considered as interfering nodes. In this paper, the interference degree is defined as the number of other nodes within the one-hop communication range between two nodes, then the average interference degree is:

$$\overline{IFD} = \frac{\sum_i^N (IFD_i(u, v) + IFD_{ic}(u, v))}{N - m} \tag{9}$$

where, since ordinary nodes are restricted to communicate only with cluster head nodes, $IFD_i(u, v)$ denotes all other nodes within the range of ordinary nodes, and $IFD_{ic}(u, v)$ denotes all the remaining nodes within the communication range of the cluster head where the node is located. N is the total number of nodes, and m is the number of sub-clusters.

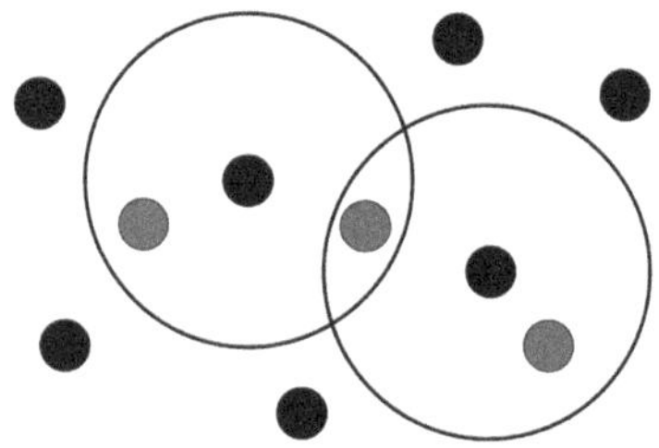

Fig. 4. Schematic calculation of interference degree

(3) Average Communication Link Length within A Cluster

For the entire communication network, the communication links between cluster head nodes are small compared to the link lengths between cluster head nodes and regular nodes, so the average intra-cluster communication link length is defined to measure the resource overhead in the network:

$$\overline{D_c} = \frac{\sum dist(u, head)}{N - m} \tag{10}$$

where *head* denotes the cluster head node in the cluster where node u is located.

3 Algorithm Design

3.1 Calculation of the Optimal Number of Clusters

For a clusterized UAV node communication network, there exists an optimal value for the number of cluster heads. If this value is small, the number of common nodes becomes relatively large and the intra-cluster communication traffic becomes large, which can lead to information congestion. Conversely, the communication load borne by the cluster heads in the backbone network becomes larger. Therefore, an optimal number of clusters should be found so that the load in the communication network is relatively balanced.

Let the communication bandwidth of ordinary nodes in the network be W_1, the communication bandwidth between cluster head nodes be W_2, and the optimal number of clusters be m. According to the literature [9, 10], the throughput of ordinary nodes within the clusters and the inter-cluster head throughput are respectively:

$$R_{local} = \theta\left(\frac{W_1}{\sqrt{N/m}}\right) \tag{11}$$

$$R_{backbone} = \theta\left(\frac{W_2}{\sqrt{m}}\right) \tag{12}$$

Assuming that the density of information and the distribution of nodes in the network are uniform, at this point the communication traffic from the cluster head node to the ordinary nodes in its cluster accounts for the overall $1/m$. To ensure that the traffic in the cluster head is not congested, there should be:

$$\left(1 - \frac{1}{m}\right)R_{local} \le R_{backbone} \tag{13}$$

To optimize the performance of the whole network, let the equal sign hold when $m = M$. Substituting Eqs. (11) (12) into Eq. (13), we obtain:

$$M = \frac{W_2}{W_1}\sqrt{N} + 1 \tag{14}$$

Assuming consistent communication bandwidth for each node. We have $W_1 = W_2$. So the optimal number of clusters is:

$$m = \sqrt{N} + 1 \tag{15}$$

3.2 Design of the Algorithm

In this paper, an improved clustering algorithm based on the fuzzy C-mean clustering algorithm is used to compare with the widely used LEACH algorithm and to draw relevant conclusions under set performance metrics. The algorithm is implemented as follows.

Let there be a total of N objects to be classified and the number of classes is c. Create an affiliation matrix to represent the magnitude of the likelihood that each object to be classified belongs to each class, and the matrix elements are denoted as μ_{ij}, where $i \in [1, c], j \in [1, n]$. Where the size of μ_{ij} is negatively correlated with the distance between the node and the clustering center:

$$\mu_{ij} = \beta_{ij} e^{-dist} \tag{16}$$

where β_{ij} satisfies:

$$\sum_{i=1}^{c} \mu_{ij} = 1 \tag{17}$$

The objective function used in the algorithm to determine if the clustering is complete is as follows:

$$J = \sum_{i=1}^{c} \sum_{j=1}^{n} {\mu_{ij}}^m \|x_j - c_i\|^2 \tag{18}$$

m is the affiliation index to influence the weighted values. To minimize the objective function, Eq. (17) as a constraint and solved by Lagrangian polarization method:

$$J = \sum_{i=1}^{c} \sum_{j=1}^{n} {\mu_{ij}}^m \|x_j - c_i\|^2 + \lambda_1 \left(\sum_{i=1}^{c} \mu_{i1} - 1 \right) + \cdots + \lambda_n \left(\sum_{i=1}^{c} \mu_{in} - 1 \right) \tag{19}$$

The above equation takes a partial derivative for μ_{ij}:

$$\frac{\partial J}{\partial \mu_{ij}} = m \|x_j - c_i\|^2 {\mu_{ij}}^{m-1} + \lambda_j = 0 \tag{20}$$

So we have:

$$\frac{1}{\mu_{ij}} = \sum_{k=1}^{c} \left(\frac{\|x_j - c_i\|}{\|x_j - c_k\|} \right)^{\frac{2}{m-1}} \tag{21}$$

The objective function is then partialized with respect to the clustering centers:

$$\frac{\partial J}{\partial c_i} = \sum_{j=1}^{n} -2{\mu_{ij}}^m (x_j - c_i) = 0 \tag{22}$$

It is solved as:

$$c_i = \frac{\sum_{j=1}^{n} \left({\mu_{ij}}^m x_j \right)}{\sum_{j=1}^{n} {\mu_{ij}}^m} \tag{23}$$

So far, in this paper, after initializing the affiliation matrix of an object to be classified, based on the iterative formula (21) (23), the position of the current clustering center c_i can be calculated, and then this iteration of the next round of affiliation matrix $\boldsymbol{\mu_{ij}}$, and so on, until it meets the set end conditions.

4 Simulation

In this section, the comparison of results between the improved clustering algorithm based on fuzzy C-clustering algorithm (FCM) and the widely used LEACH algorithm will be investigated along with the design of the relevant parameters and final conclusions will be drawn under the collaborative work of UAV clusters.

In this paper, we preset the size of the site to be $4000 \times 4000\ m^2$, and place N UAV nodes in the area, in which the limiting distance R_{max} for communication between ordinary nodes and cluster head nodes are all the same, and the cluster head nodes are mutually reachable. The movement of cluster head nodes conforms to the random wandering model and the movement of ordinary nodes conforms to the group pursuit model. The parameters are set in the Table 1:

Table 1. Initial simulation Parameter Setting

simulation parameter	value
bounds	$3000 \times 3000\ m^2$
Number of nodes	[50,150]
Maximum communication distance	600 m
Moving speed range	[20,40] m/s
Running time	[20,50] s
Time step	0.25 s
iterations	200
Subjection degree	2

The positions of 81 nodes are initialized in a preset range and the simulation parameters are given in Table 2.

Table 2. Algorithm simulation Parameter Setting

simulation parameter	value
Number of nodes	81
Running time	20 s
Moving speed	20–40 m/s
Grouping number	10

According to the discussion in this paper, the optimal number of clusters should be $M = \sqrt{N} + 1 = 10$, then the LEACH algorithm and the initial clustering results of the algorithm used in this paper are respectively (Figs. 5 and 6):

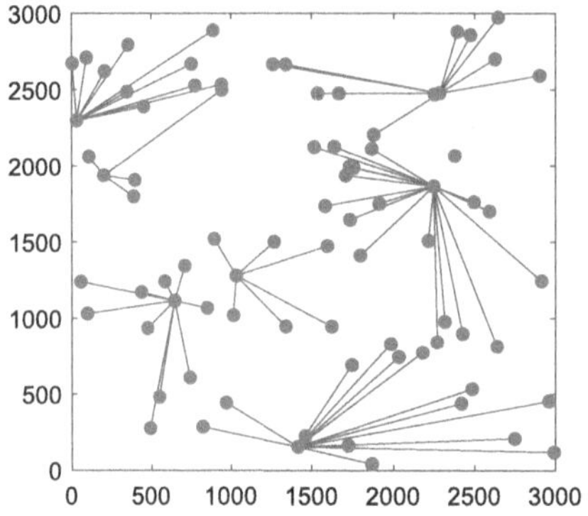

Fig. 5. LEACH initial topological graph

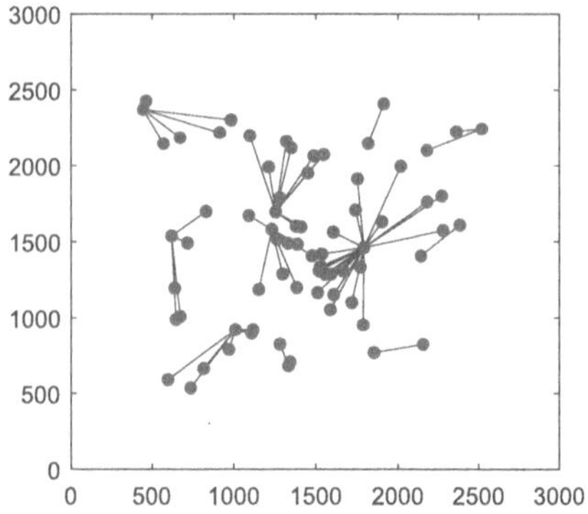

Fig. 6. LEACH final topological graph

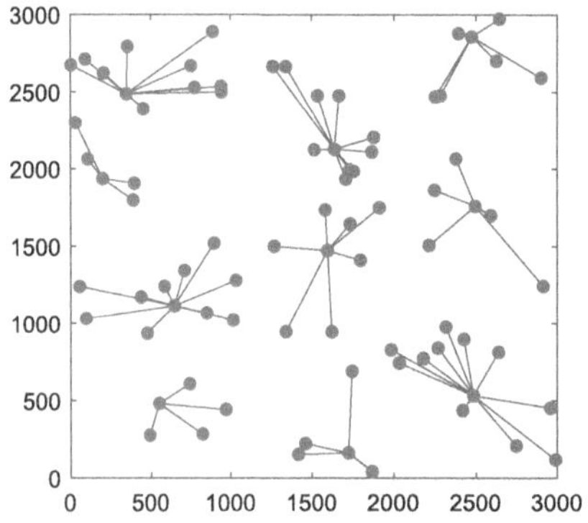

Fig. 7. FCM initial topological graph

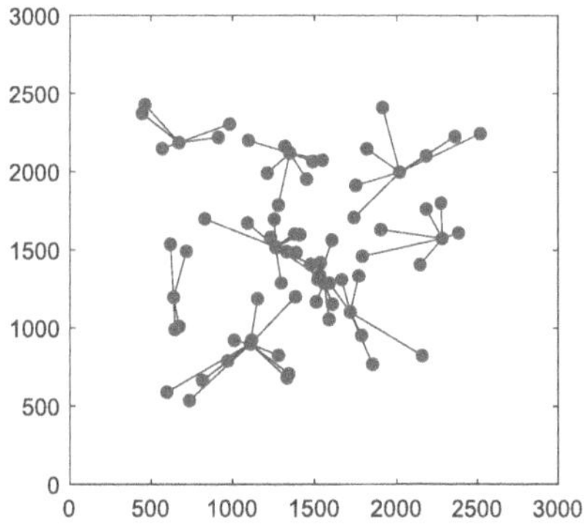

Fig. 8. FCM final topological graph

Next the position matrix of the nodes is obtained by letting the normal nodes of the UAV follow the group pursuit model and the cluster head nodes follow the random wandering model. After running for a preset simulation time, the final topology results obtained by both methods are shown in Fig. 7 and Fig. 8:

It can be seen that compared with the clustering scheme of LEACH algorithm, the improved clustering algorithm based on fuzzy C-clustering used in this paper has more uniform and beautiful clustering results. And the latter has fewer inter-cluster transfers occurring during the iteration of the algorithm.

The comparison of the relevant metrics of the two algorithms is shown in Fig. 9:

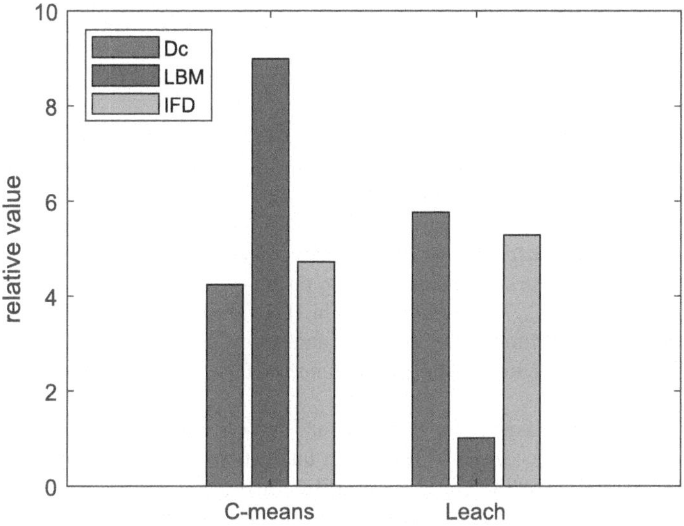

Fig. 9. Performance comparison of the two algorithms

It can be seen that for load balancing degree, the algorithm used in this paper is much higher than LEACH algorithm. For average path length and average interference, the algorithm used in this paper is lower than LEACH algorithm and shows better performance.

5 Conclusion

In this paper, an improved clustering algorithm based on fuzzy C clustering algorithm is proposed for UAV cluster cooperative work scenario. Firstly, the mobile model of UAVs is modeled, then the principle of this paper's algorithm is analyzed and several evaluation indexes are proposed, and finally the two algorithms are simulated and analyzed. The experimental results show that the algorithm incorporates the concept of affiliation, which is negatively correlated with the Euclidean distance between nodes, and has better performance than the widely used LEACH algorithm. Using this algorithm can show better topology control results when performing UAV cluster cooperative work.

Acknowledgement. This work has received support from the National Natural Science Foundation of China under Grant 62171158 and National Key Laboratory of Air-based Information Perception and Fusion and Aeronautical Science Foundation of China under Grant 20230001004002 and it was also supported by the Fundamental Research Funds for the Central Universities under Grant 2242022k60006.

References

1. Cheng, X., Dong, C., Dai, H., Chen, G.: MOOC: a mobility control based clustering scheme for area coverage in FANETs. In: 2018 IEEE 19th International Symposium on A World of Wireless, Mobile and Multimedia Networks (WoWMoM), pp. 14–22. IEEE (2018)

2. Lin, Z., Liu, H.H., Wotton, M.: Kalman filter-based large-scale wildfire monitoring with a system of UAVs. IEEE Trans. Ind. Electron. **66**(1), 606–615 (2018)
3. Prim, R.C.: Shortest connection networks and some generalizations. Bell Syst. Tech. J. **36**(6), 1389–1401 (1957)
4. Abiuzi, L.B., Cesar, C.D.A., Ribeiro, C.H.: A-lmst: an adaptive LMST local topology control algorithm for mobile ad hoc networks. In: 2016 IEEE 41st Conference on Local Computer Networks (LCN) pp. 168–171. IEEE (2016)
5. Tang, J., Xue, G., Zhang, W.: Interference-aware topology control and QoS routing in multi-channel wireless mesh networks. In: Proceedings of the 6th ACM International symposium on Mobile ad hoc Networking and Computing, pp. 68–77 (2005)
6. Biradar, A., Thool, R.C., Velur, R., Indumathi, T.S.: Dual channel based multi-objectives genetic routing protocol for ad-hoc networks and optical networks using power aware clustered topology. In: 2012 International Conference on Optical Engineering (ICOE), pp. 1–6. IEEE (2012)
7. Khoufi, I., Minet, P., Laouiti, A.: Fault-tolerant and constrained relay node placement in wireless sensor networks. In: 2016 IEEE 13th International Conference on Mobile Ad Hoc and Sensor Systems (MASS), pp. 127–135. IEEE (2016)
8. Li, J., et al.: Centroid-guided target-driven topology control method for UAV ad-hoc networks based on tiny deep reinforcement learning algorithm. IEEE Internet Things J. (2024)
9. Gupta, P., Kumar, P.R.: The capacity of wireless networks. IEEE Trans. Inf. Theory **46**(2), 388–404 (2000)
10. Liu, D., Xu, Y., Shen, L., Xu, Y.: Self-organising multiuser matching in cellular networks: a score-based mutually beneficial approach. IET Commun. **10**(15), 1928–1937 (2016)

Research on Joint Optimization Model of Drone Trajectory and Power for SAGIN Swarm Intelligence Technology

Zhenyu Xu(✉), XiuLian Lin, and XinLu Li

Huizhou Engineering Vocational College, Huizhou 516023, Guang Dong, China
hitusa@126.com

Abstract. To revitalize rural areas, technology must take the lead. Only by modernizing agriculture can China achieve complete modernization. In recent years, SAGIN (Space-Air-Ground Integrated Network) has received widespread attention from academia and industry, and can be applied in many practical fields such as global surveying, intelligent transportation systems, military missions, and more. As an information carrier for aerial platforms, the deployment strategy of agricultural and forestry plant protection drones is crucial to the performance of the system. This paper mainly studies how to establish a mathematical model for maximizing energy efficiency of unmanned aerial vehicles (UAVs) based on SAGIN swarm intelligence technology, under the constraints of energy and user rate, by jointly optimizing launch power and flight trajectory.

Keywords: Joint Optimization Model · Drone · SAGIN · Swarm Intelligence

1 Background

The implementation of the rural revitalization strategy represents a significant decision and deployment by the Party Central Committee regarding the work related to agriculture, rural areas, and farmers (the "three rural issues"), serving as the overarching approach for addressing the "three rural issues" in the new era. The Law of the People's Republic of China on Promoting Rural Revitalization came into effect on June 1, 2021. Only when decisive progress is made in rural revitalization and agricultural and rural modernization is basically achieved can China fully realize modernization.

As a major agricultural country, China's foremost livelihood concern is to address the food and clothing needs of its 1.4 billion people. Ensuring food security and upholding the red line of 1.8 billion acres of arable land are fundamental national policies and strategies. With the rapid development of China's industry and the accelerating urbanization process, an increasing number of people are reluctant to settle in "the yellow earth," and long-standing issues such as limited per capita arable land and scattered, unconcentrated farmland have become increasingly prominent. China's agricultural development cannot afford to slacken for a moment. Therefore, the annual "No. 1 Document" of the state

X. Li et al. (Eds.): 6GN 2024, LNICST 615, pp. 125–133, 2025.
https://doi.org/10.1007/978-3-031-93193-2_10

focuses on agriculture and rural issues. On February 3, 2024, the CPC Central Committee and the State Council issued the "Opinions on Learning from the Experience of the 'Thousand Villages Demonstration, Ten Thousand Villages Renovation' Project to Effectively Advance Comprehensive Rural Revitalization." This is the 12th Central No. 1 Document guiding the work related to the "three rural issues" since the 18th National Congress of the Communist Party of China. The document emphasizes optimizing the strategic layout of agricultural science and technology innovation, supporting the construction of major innovation platforms, intensifying efforts in core technology research, and enhancing the overall efficiency of the agricultural science and technology innovation system. To achieve this, a fundamental requirement is the widespread application of new agricultural machinery and equipment. Research and promotion efforts related to these technologies are in full swing.

Agricultural and forestry plant protection work is crucial to ensuring the yield and quality of crops and economic plants. Currently, there are three primary forms of agricultural and forestry plant protection operations: manual operations, ground machinery operations, and aerial operations. Manual operations are inefficient and pose safety risks to operators, and will gradually be replaced by the latter two. Ground machinery operations, while effective, have limitations in terms of low utilization of pesticides, relatively high costs, and potential damage to crop stems, leaves, and soil structures, leading to a gradual decline in market share over the next few decades. Aerial operations, primarily utilizing rotary-wing aircraft, helicopters, large fixed-wing aircraft, and other emerging intelligent agricultural machinery and equipment to spray over crops, offer high operational efficiency and effective pest control, serving as a powerful driving force for the future of intelligent, precise, and large-scale agricultural plant protection. Plant protection drones, as the most cutting-edge high-tech in the current agricultural plant protection industry, have quickly attracted widespread attention in the agricultural plant protection market due to their unique technological advantages and operational characteristics. The emergence of plant protection drone technology has facilitated the monitoring and control of pests and diseases, investigation of agricultural and forestry resources, monitoring of agricultural and forestry fires, and monitoring of wildlife in forest areas. While improving operational efficiency and reducing manual workload, it also enhances real-time monitoring and control capabilities for agriculture and forestry, and enhances emergency response capabilities. Therefore, in China, whether in the south or north, hills or plains, small-scale or large-scale operations, the addition of plant protection drone technology will play an important role and have broad application prospects in the future agricultural plant protection market.

2 Advancements and Future Prospects of UAVs in Agricultural Plant Protection

The emergence of unmanned aerial vehicles (UAVs) in agricultural plant protection, as the cutting-edge technology at the forefront of the industry, has garnered widespread attention in the market due to their unique technological advantages and operational characteristics. The introduction of UAV technology has facilitated pest monitoring and control in forests, agricultural and forestry resource surveys, fire monitoring, wildlife monitoring, and other agroforestry tasks. It not only enhances operational efficiency and reduces manual workload but also improves real-time monitoring capabilities and emergency response speed for agroforestry. Consequently, the integration of UAV technology plays a pivotal role in agricultural plant protection, regardless of location (south or north China, hills or plains), and scale (small or large-scale operations), with promising application prospects in the future market.

Currently, agricultural plant protection primarily comprises three operational forms: manual operations, ground-based machinery operations, and aerial operations. Manual operations, due to their inefficiency and safety hazards, are gradually being phased out. Meanwhile, ground-based machinery, while effective, is costly and prone to damaging crops, leading to a gradual decline in market share. In contrast, aerial operations, particularly those utilizing UAVs, are emerging as the dominant approach in future agricultural plant protection due to their high operational efficiency and effective control outcomes.

The application of UAVs in agricultural plant protection boasts several significant advantages.(1)High operational efficiency: UAVs can swiftly cover vast agricultural areas, leading to substantial improvements in work efficiency.(2)Effective control outcomes: UAVs enable precise pesticide application, reducing waste and enhancing control effectiveness.(3)Reduced labor costs: By minimizing the reliance on manual labor, UAVs alleviate labor intensity and reduce costs.(4)Enhanced real-time monitoring capabilities: Equipped with various sensors, UAVs can continuously monitor agroforestry conditions, bolstering emergency response capabilities.

With the acceleration of industrialization and urbanization, agricultural plant protection faces new challenges that necessitate more efficient and intelligent solutions. UAVs, with their distinctive benefits, are poised to revolutionize this field and contribute significantly to the modernization of agricultural practices.

Foreign countries initiated research into unmanned aerial vehicles (UAVs) for agricultural plant protection at an earlier stage. As early as 1911, the Germans first proposed the idea of using airplanes to spray pesticides to control agricultural and forestry pests. In 1912, Canada had already started utilizing fixed-wing aircraft as auxiliary tools for pesticide spraying on large field crops and orchards. In 1918, the United States officially kicked off the era of aerial plant protection by successfully using airplanes to spray pesticides against locusts. By 1987, Japan's Yamaha Company produced the world's first agricultural plant protection UAV (R-50), which was widely applied in pest control operations for large field crops. After three decades of intensive research and development, Japan gradually emerged as the world's leading country in the use of UAVs for pesticide spraying.

Within China, the advancement of agricultural modernization has spurred rapid growth in UAV technology for agricultural plant protection. An increasing number of agricultural enterprises and research institutions have invested in the research and application of agricultural plant protection UAVs, driving the process of intelligent, precise, and large-scale agricultural plant protection forward. This technology not only enhances operational efficiency and reduces manual labor but also improves real-time monitoring capabilities and emergency response in agricultural and forestry industries. From forest pest and disease monitoring to resource surveys, fire monitoring, and wildlife monitoring, agricultural plant protection UAVs have garnered widespread attention in the market due to their unique technical advantages and operational characteristics.

In recent years, SAGIN (Space-Air-Ground Integrated Network), as an emerging network architecture, has garnered significant attention from both academia and industry (Wei C, Zhang Y, Wang R, et al., 2021) [1]. SAGIN finds extensive applications across various domains, including global surveying and mapping, intelligent transportation systems, military missions, and more. The issue of drone path planning has long been a research hotspot, particularly when drones collaborate with ground transportation tools (Jeong H Y, Lee S., 2023) [2]. This collaborative working mode is of great significance for enhancing the operational efficiency and coverage of agricultural plant protection drones, which is also a focus of this study. With the advent of the 6G era, SAGIN has demonstrated tremendous potential in intelligent transmission, particularly through technological means such as federated learning (Tang F, Wen C, Kato C N., 2023) [3]. This intelligent transmission capability is crucial for improving the energy efficiency of agricultural plant protection drones. To fully comprehend the application of SAGIN in 6G, it is essential to consider its architecture, requirements, and associated challenges (Cui H, Zhang J, Geng Y, et al., 2022) [4]. These elements are vital for the deployment and optimization of drones in the field of agricultural plant protection within this study. In SAGIN, the rational deployment of multiple controllers is crucial for enhancing network scalability and flexibility (Liao Z, Chen C, Ju Y, et al., 2022) [5]. This research finding offers valuable insights into the management and control strategies of drone networks in our study. Researchers have conducted extensive work on the performance analysis of SAGIN in complex environments, including performance evaluations over arbitrarily correlated multivariate free-space optics (FSO) channels (Alimi, Isiaka, A, et al., 2018) [6]. These research outcomes provide important references for assessing the communication performance of drones in agricultural plant protection tasks. Intelligent access technology represents a crucial aspect of SAGIN networks for achieving efficient communication and management (Liu L, Liu Q., 2022) [7]. Through intelligent access technology, drones can more flexibly connect to the SAGIN network, enabling rapid data transmission and processing, which is significant for real-time monitoring and control of drones in agricultural plant protection tasks within this study. Addressing the issue of drone scheduling, researchers have proposed various optimization algorithms, such as particle swarm optimization algorithms optimized by deep reinforcement learning (Zai W, Wang J, Li G, et al., 2023) [8]. These algorithms offer new perspectives for efficient drone scheduling in complex environments, providing valuable insights for optimizing drone scheduling in agricultural plant protection tasks within this study. To optimize the deployment of service function chains in SAGIN and enhance system efficiency, researchers have proposed deployment algorithms based on deep reinforcement learning (Feng X, He M, Zhuang Y P R., 2024) [9]. This algorithmic approach offers inspiration for the joint optimization of drone trajectories and power within this study. To improve the energy efficiency of drones in emergency communications, researchers have proposed multiple drone deployment schemes (Zhang S, Shi S, Wang W, et al., 2022) [10]. However, most of these schemes are tailored to specific scenarios, lacking universality and flexibility. This study aims to achieve high-efficiency operation of agricultural plant protection drones in the SAGIN network through the joint optimization of drone trajectories and power.

3 System Model and Problem Definition

This thesis primarily focuses on establishing a mathematical model for maximizing energy efficiency through the joint optimization of transmission power and flight trajectory of unmanned aerial vehicles (UAVs) leveraging software-defined air-ground integrated network (SAGIN) swarm intelligence technology, under constraints related to energy and user rates. Here, UAVs serve as aerial base stations, covering terrestrial users (TUs) distributed within the target region according to a Poisson point process with a density denoted as "λ". Given that a single UAV cannot achieve full coverage of the target area, continuous deployment of UAVs is necessary. Furthermore, we assume that TUs remain stationary until the UAVs complete their coverage, implying that the coordinates of TUs do not change over time. Additionally, we assume that UAVs possess sufficient bandwidth resources to allocate orthogonal channels to all users.

Drawing upon the correlation between swarm control theory and UAV formation flight, this paper proposes a distributed swarm control approach to facilitate efficient formation and autonomous networking under large-scale unmanned platforms. Specifically, based on the overall distribution of nodes, all nodes are classified into two categories: ordinary nodes and swarm leaders. The differences in energy resources, storage, and computational capabilities between swarm leaders and ordinary nodes are minimal, with the primary selection factors being their positions in the network and the types of tasks they execute. This facilitates flexible replacement of swarm leaders, supporting highly dynamic network topologies. Both swarm leaders and the ground control center, as well as ordinary nodes and swarm leaders, form star-shaped networks. When tasks are executed, swarm leaders operate in search mode, performing detection and information collection tasks, while ordinary nodes search for nearby swarm leaders based on the speed and location information of surrounding nodes and follow them. A swarm leader, along with several ordinary nodes (depending on task requirements), forms a swarm, which conducts autonomous networking and completes tasks collaboratively. The following sections provide detailed descriptions in terms of system model, mode control, and information exchange.

System model: Consider a network of N nodes, and the state of one node can be described by (1)

$$\begin{cases} \dot{q}_i(t) = p_i(t) \\ \dot{p}_i(t) = u_i(t),\ i = 1, 2, \ldots, N, \end{cases} \tag{1}$$

$p_i(t)$ and $q_i(t)$ represent the speed and position of node i at time t, respectively, and $u_i(t)$ represents the control information input of node i at time t. When a node moves within a region, define the set of neighboring nodes according to (2).

$$N_i(t) = \{j : \left\| q_j - q_i \right\| < r_c, j = 1, 2, \ldots, N, j \neq i\} \tag{2}$$

Node i records the actions and positions of the group leader node through its own information list. Use l_i to represent the information list of node i, assuming that a group leader k moves at a speed of p_k^n and an acceleration of a_k^n at time t at q_k^n. If a regular node discovers a group leader near it, it records the corresponding information in its own information table, as shown in (3).

$$l_i(k) = \left\{k, q_k^n, p_k^n, a_k^n, t\right\} \tag{3}$$

If the node itself cannot obtain information about the group leader, it exchanges a list of information with its neighboring nodes within one hop range. If the group leader is not yet perceived within one hop range, it is decided to enter a sleep state based on the distribution of surrounding nodes to reduce energy resource consumption or become a new group leader centered around the node.

Definition n_{max} is the maximum number of groups including the group leader, and n_{ik} is the number of nodes following the group leader k at time t.

The system model considered in this paper consists of 1 drones and M TUs. In this paper, the trajectory of UAV base stations is discretized into several stopping points.

In a two-dimensional Cartesian coordinate system, the coordinates of the drone at the nth SP can be expressed as $\mathbf{q}_n = \left(x_n^n, y_n^n\right) n \in \mathbb{N} \triangleq \{1,2, \ldots, N\}$ where N is the total number of stopping points.

Due to the limitation of battery power in the coverage area of the drone, we assume that the number of TUs covered by the drone at the nth SP is K_n, $\sum_{n=1}^{N} K_n = M$. Then, the two-dimensional coordinates of the kth TU covered by the drone at the nth SP are represented as $\mathbf{W}_n^k = \left(x_n^k, y_n^k\right) \in \mathbb{W}_n$, where W_n is the coordinate set of K_n TU, $\mathbf{W}_n^k = \left(x_n^k, y_n^k\right) \in \mathbb{W}_n$. According to the Euclidean formula, the distance between the nth SP and the kth TU of the drone is expressed as:

$$d_n^k = \sqrt{\left\| q_n - w_n^k \right\|^2 + H^2} \tag{4}$$

Generally, the LoS channel is superior to other channels in the UAV-TU wireless communication link. We consider the wireless channel of the drone at the nth SP as a free space path loss model, which can be expressed as

$$h_n^k = \beta_0 \left(d_n^k\right)^{-\alpha k}_{n} \tag{5}$$

where β_0 is the channel power gain at a reference distance of 1m, and α is the path loss index. Then apply Shannon capacity to define the rate of the kth TU as

$$R_n^k = B\log_2\left(1 + \frac{\gamma_n^k}{\left(d_n^k\right)^\alpha}\right) \tag{6}$$

where B is the bandwidth, and $\gamma_n^k \triangleq \frac{P_n^k \beta_0}{\sigma^2}$ is the received signal-to-noise ratio at a reference distance of 1m. $P_{\mathrm{n}}^{\mathrm{k}}$ is the transmission power of the kth TU of the drone at the nth SP, and σ^2 is the power of additive white Gaussian noise.

This paper assumes that the drone completes the closed-loop coverage task at a constant speed v, i.e. hovers N times and returns to the starting point. The total flight energy consumption of drones is expressed as

$$E^f = \sum\nolimits_{n=1}^{N+1} c_f l_n \tag{7}$$

where l_n represents the distance between adjacent SPs, $c_f = \left(c_1 v^2 + \frac{c_2}{v^2}\right)$, where c_I and c_2 are two parameters related to drone weight, wing area, and air density. Specifically, the relationship between the SP and TU of the drone is shown in Fig. 1.

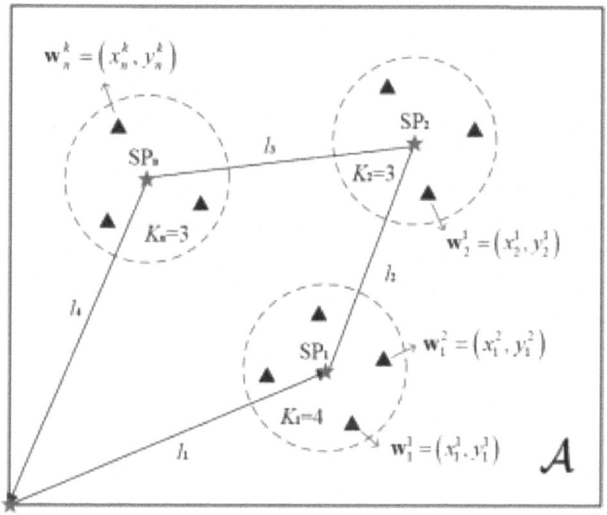

Fig. 1. Schematic diagram of the deployment location of drone docking points

Then, the total energy efficiency of the drone can be calculated as

$$\mathbf{E} = \frac{\sum_{n=1}^{N} \sum_{k=1}^{K_n} R_n^k}{\sum_{n=1}^{N} \sum_{k=1}^{K_n} P_n^k + E^f} \tag{8}$$

In order to improve the energy efficiency of drones, the goal of this article is to jointly optimize deployment positions and power allocation. Therefore, the problem of maximizing energy efficiency of agricultural and forestry plant protection drones is expressed as follows

$$\begin{aligned}
&\mathrm{P} : \max_{\eta, X'} \mathrm{E}' \\
&\text{s.t.}\, \mathrm{C}_1 : \log\left(\frac{1}{\psi}\right) + \log\left(\mathrm{r}_{\mathrm{n}}^{\mathrm{k,t}} + \mathrm{H}^2\right) - \tilde{\mathrm{P}}_{\mathrm{n}}^{\mathrm{k}} \le 0, \forall \mathrm{n}, \forall \mathrm{k}, \\
&\mathrm{C}_2 : \sum\nolimits_{\mathrm{k}=1}^{\mathrm{K_n}} \mathrm{e}^{\tilde{\mathrm{P}}_{\mathrm{n}}^{\mathrm{k}}} \le \mathrm{P}_{\mathrm{sum}}, \forall \mathrm{n}, \\
&\mathrm{C}_3 : \mathrm{c_f} \sum\nolimits_{\mathrm{n}=1}^{\mathrm{N}+1} \mathrm{l_n} \le \mathrm{E}_{\mathrm{f}}^{\max}, \\
&\mathrm{C}_4 : \mathrm{r}_{\mathrm{n}}^{\mathrm{c}} = \left\|\mathrm{q_n} - \mathrm{w}_{\mathrm{n}}^{\mathrm{c}}\right\|^2 \le \left(\mathrm{r}_{\min}^{\mathrm{n}}\right)^2, \forall \mathrm{n}.
\end{aligned} \tag{9}$$

Among them, $\mathbf{P} = \{P_n^k, \forall n, k\}$ and $\mathbf{Q} = \{\mathbf{q}_n, \forall n\}$. Under the above constraints, *C1* and *C2* ensure the minimum rate of each TU and the power limitation of UAV-BS, respectively. *C3* represents the constraint on the energy consumption of drone flight.

4 Conclusion

The proposed model, by optimizing the flight trajectory and power allocation of unmanned aerial vehicles (UAVs), significantly enhances the energy efficiency of the UAVs, thereby reducing energy consumption and operational time. Furthermore, the efficient operation of UAVs contributes to the improvement of crop yield and quality, safeguarding national food security.

While this model has made considerable progress in the joint optimization of UAV trajectories and power, there are still numerous issues worthy of further investigation. Notably, the nonlinear coupling between the UAV's position coordinates and transmit power results in a non-convex objective function. Additionally, the parameters N and Kn need to be determined prior to solving the optimization problem P1. Consequently, due to the non-convexity of the objective function and constraints, as well as the presence of undetermined parameters, P1 poses a challenging problem that requires continued research to develop suitable algorithms. Ultimately, the aim is to find an optimization scheme that outperforms existing approaches in terms of both energy efficiency and task completion time for agricultural plant protection UAVs.

References

1. Wei, C., Zhang, Y., Wang, R., et al.: Key technologies of space-air-ground integrated network: a comprehensive review. Mob. Multimed. Commun. (2021). https://doi.org/10.1007/978-3-030-89814-4_5
2. Jeong, H.Y., Lee, S.: Drone routing problem with truck: optimization and quantitative analysis. Expert Syst. Appl. (Oct.), 227 (2023)
3. Tang, F., Wen, C., Kato, C.N.: Federated learning for intelligent transmission with space-air-ground integrated network toward 6G. IEEE Netw. Mag. Comput. Commun. **37**(2), 198–204 (2023)
4. Cui, H, et al.: Space-air-ground integrated network (SAGIN) for 6G: requirements, architecture and challenges. China Commun. **19**(2), 90–108 (2022)
5. Liao, Z., Chen, C., Ju, Y., et al.: Multi-controller deployment in SDN-enabled 6G space-air-ground integrated network. Remote. Sens. **14**, 1076 (2022). https://doi.org/10.3390/rs14051076
6. Alimi, I.A., et al.: Performance analysis of space-air-ground integrated network (SAGIN) over an arbitrarily correlated multivariate FSO channel. Wirel. Pers. Commun. Int. J. (2018)
7. Liu, L., Liu, Q.: Research on intelligent access of space-air-ground integrated network. In: Sun, S., Hong, T., Yu, P., Zou, J. (eds.) Signal and Information Processing, Networking and Computers. ICSINC 2021. LNEE, vol. 895, pp. 66–79. Springer, Singapore (2022). https://doi.org/10.1007/978-981-19-4775-9_9
8. Zai W , Wang J , Li G ,et al.A Drone Scheduling Method for Emergency Power Material Transportation Based on Deep Reinforcement Learning Optimized PSO Algorithm[J]. 2023

9. Feng, X., He, M., Zhuang, Y.P.R.: Service function chain deployment algorithm based on deep reinforcement learning in space-air-ground integrated network. Future Internet **16**(1) (2024)
10. Zhang, S., Shi, S., Wang, W., et al.: An energy-efficient UAV deployment scheme for emergency communications in air-ground networks with joint trajectory and power optimization. China commun. **19**(7), 12 (2022). https://doi.org/10.23919/JCC.2022.07.006

Research on the Application of LoRa Technology in Intelligent Ranch Environmental Monitoring System

Yanhong Zhang, Zhenyu Xu(✉), and Jingzhong Li

Huizhou Engineering Vocational College, Huizhou 516023, Guangdong, China
hitusa@126.com

Abstract. This paper extensively examines the integration of LoRa technology into smart livestock farm environmental monitoring systems, aiming to elevate the informatization level in the livestock industry through modern technological advancements. It initially outlines the current state of China's livestock industry and the pressing need for modernization. Subsequently, the principles and advantages of LPWAN technology, particularly LoRa, are introduced. As an ultra-long-range wireless transmission solution, LoRa boasts low power consumption, long-range transmission capabilities, and massive connectivity, offering technical support for the intellectualization of the livestock sector. This paper not only analyzes LoRa technology's working principles, spread spectrum communication principles, and data packet formats but also elaborates on the design, configuration, and implementation process of a smart livestock farm environmental monitoring system based on LoRa. Additionally, it explores specific challenges encountered during implementation and their respective solutions. By leveraging LoRa technology, this paper underscores its potential to enhance the informatization of livestock farms, facilitating the collection and transmission of environmental data crucial for effective monitoring and management. The discussion highlights LoRa's robustness in tackling the complexities of livestock farm environments, addressing challenges posed by such environments, and providing practical solutions for overcoming them. This academic endeavor contributes to advancing the smart and information-driven transformation of the livestock industry.

Keywords: LoRa Technology · Smart Ranch · Digital Transformation of Agriculture

1 Introduction

1.1 Background

China is a prominent animal husbandry country, consistently ranking among the top globally in terms of total animal husbandry production. The development of animal husbandry necessitates attention to the modernization of pastures. The integration, mechanization,

X. Li et al. (Eds.): 6GN 2024, LNICST 615, pp. 134–142, 2025.
https://doi.org/10.1007/978-3-031-93193-2_11

and informatization of animal husbandry have emerged as pivotal foci of digital transformation within the industry. The 14th Five-Year Plan for Promoting Agricultural and Rural Modernization advocates the development of modern animal husbandry, guiding large-scale meat, beef, and sheep farms towards standardized, intensive, and intelligent transformation of livestock and poultry pens. Therefore, enhancing the informatization level of pastures has become a critical factor in China's vigorous development of animal husbandry and breeding. In recent years, with the continuous advancement of science and technology, emerging technologies such as the Internet of Things (IoT), big data, and artificial intelligence have driven the transformation of traditional animal husbandry methods towards smart animal husbandry. In response to the urgent need for modernization in China's livestock industry, this paper, grounded in the exploration of LoRa technology, undertakes a detailed analysis of its practical application in pasture environment monitoring. Furthermore, it delves into potential challenges encountered during implementation and proposes comprehensive solutions. This academic endeavor aims to contribute to the informed adoption of LoRa technology in livestock farms, ensuring an effective and sustainable transition towards smarter and more information-driven operations.

1.2 Overview of LoRa Technology

LPWAN technology is characterized by low bandwidth, long distance, low power consumption, and a large number of connections. The communication technologies employed can be broadly categorized into two groups based on their operating frequency bands: those operating on unauthorized spectrums, such as LoRa (Long Range) and SigFox, and those supported by the 3GPP protocol and operating on operator-authorized spectrums, including EC-GSM, NB-IoT, among others. The latter's construction necessitates the development based on 2/3/4G cellular communication networks.

The LoRaWAN standard, promoted by the LoRa Alliance, has enabled secure, long-range communication for IoT applications, significantly contributing to smart environmental monitoring, as exemplified in Kannayeram et al.'s work (2022). [1] LoRa technology has found numerous applications in various fields, including but not limited to, tobacco storage, as demonstrated in Tang et al.'s study (2022), where LoRa was utilized for monitoring the environment of tobacco storage facilities. [2] In the context of controlled environments like greenhouses, LoRa-based systems have demonstrated their effectiveness in environmental monitoring and control, as outlined in Ou et al.'s design (2021). [3] For indoor environments, where precise monitoring is equally crucial, LoRa-based systems have been employed effectively, as evidenced by Jiang et al.'s indoor environment quality monitoring system (2022). [4] For specialized environmental monitoring tasks such as wind monitoring, LoRa technology has proven to be an effective solution, as shown in the development of wind monitoring systems by Purnomo et al. (2022). [5] Water quality monitoring, a crucial aspect of environmental surveillance, has benefited significantly from IoT-based systems employing LoRa technology, as evidenced by Promput et al.'s research (2023). [6] The integration of LoRa technology in aquaculture has led to the development of intelligent environment monitoring systems, as demonstrated in the work by Huiying et al. (2021), highlighting its potential in diverse agricultural sectors. [7] In precision agriculture, LoRa-based wireless sensor networks have shown great potential in energy-efficient environmental monitoring, as highlighted by Krianovi et al. (2023). [8] Beyond traditional agricultural applications, LoRa technology has also been successfully deployed in urban smart management systems, such as the LoRa-based garbage monitoring system developed by Kumar et al. (2022) [9].

This paper integrates LPWAN technology to design a ranch environment monitoring system based on LoRa amidst the rapid development of the IoT and examines the information transmission methods of multiple terminal nodes in outdoor environments utilizing LoRa.

While LoRa technology boasts numerous advantages, its implementation in pasture environments poses various challenges due to the intricate and dynamic nature of such settings. These environments encompass open fields, forests, hills, and more, all of which can potentially interfere with or attenuate LoRa signal propagation. For instance, obstructions from trees and buildings can reduce signal coverage, while variations in humidity and temperature may compromise signal quality. Addressing these environmental factors is essential to ensuring the reliability and effectiveness of LoRa-based monitoring systems in livestock farms.

2 The Principle of LoRa Technology

LoRa is an ultra-long-distance wireless transmission technology rooted in spread spectrum principles, promoted by Semtech in the United States. The LoRa Alliance, an open and non-profit organization established by Semtech in March 2015, has since grown to comprise over 500 members, making it one of the largest and fastest-growing alliances in the technology sector. Members hail from diverse global backgrounds, including multinational communication firms, equipment manufacturers, system integrators, sensor producers, startups, and semiconductor companies. Collectively, they strive to position LoRaWAN as a preeminent open global standard, enabling secure, operator-level IoT LPWAN connectivity.

LoRa operates within the unlicensed ISM (Industrial, Scientific, and Medical) frequency band, although the specific utilization of this band may vary across different countries or regions. In the Chinese market, the China LoRa Application Alliance (CLAA), led by ZTE, recommends the adoption of the 470–518 MHz frequency band for LoRa applications. Meanwhile, radio metering instruments operate within the 470–510 MHz frequency band.

LoRa operates within unlicensed frequency bands, enabling network construction without the need for prior applications, thus facilitating a simplified network architecture and lower operational costs. This technology employs linear spread spectrum modulation, which significantly enhances communication distance while maintaining low power consumption. Crucially, even when terminals with distinct spread spectrum sequences transmit on the same frequency, they do not interfere with one another. Consequently, devices built on this foundation can concurrently receive and process data from multiple nodes, substantially expanding system capacity.

LoRa gateways boast the capability to support over 20,000 terminal connections, underscoring their scalability and versatility. The coverage distance of a single LoRa gateway typically ranges from 3 to 5 km, outperforming traditional cellular networks in complex urban environments and even extending to over 15 km in open areas. Under specific conditions, such as in rural or remote locations with minimal obstructions, communication distances of up to 100 km have been successfully achieved.

The advantages of LoRa are particularly pronounced in scenarios requiring long-range, low-power communication, such as livestock farm environmental monitoring. By leveraging LoRa technology, farms can implement efficient and cost-effective monitoring systems that provide real-time data on critical environmental parameters. This, in turn, facilitates informed decision-making and enhances the overall productivity and sustainability of livestock operations.

LoRa, an asynchronous protocol based on the ALOHA mechanism, allows nodes to alternate between extended and brief sleep periods, contingent upon the specific demands of the application scenario. In its operational mode, LoRa exhibits a low signal-to-noise ratio requirement, leading to reduced signal power consumption. Consequently, battery life can extend for several to over ten years. Furthermore, the cost of LoRaWAN modules is approximately $4–5, constituting approximately half the price of comparable cellular LTE modules, such as NB-IoT.

LoRa operates within three distinct working modes: Class A, Class B, and Class C.

Class A: Initiated by the terminal, this mode permits uplink transmission followed by a brief receiving window. Consequently, the terminal can only receive data subsequent to a successful uplink transmission. This arrangement offers unconstrained uplink capabilities, with downlink reception limited to the period following an uplink packet. Notably, Class A offers the lowest power consumption among the three modes.

Class B: Here, the terminal and server engage in negotiations to determine the timing and duration of the receiving window, facilitating the simultaneous reception of multiple packets. This mode balances power consumption with flexibility, enabling lower power consumption than Class C while maintaining a degree of responsiveness.

Class C: In this continuous reception mode, the terminal maintains an open receiving window except during transmission periods. Although this approach minimizes communication delays, it necessitates higher power consumption, rendering it the most energy-intensive among the three modes.

LoRa technology, underpinned by spread spectrum communication, significantly enhances signal anti-interference capabilities and transmission distances. This robustness is particularly evident in complex pasture environments characterized by varied terrain and numerous obstacles. By optimizing receiver sensitivity and extending communication ranges, LoRa ensures stable signal transmission, thereby safeguarding the integrity and reliability of data monitoring systems in livestock farms. In these environments, the adaptability of LoRa technology is crucial for effective environmental monitoring and management.

3 Design of LoRa-Based Smart Ranch Environmental Monitoring System

3.1 System Composition

The system comprises LoRa sensor nodes, LoRa gateways, and a host computer. To tackle the intricacies of the ranch environment, we have specifically optimized the layout of sensor nodes and gateway positions to ensure comprehensive signal coverage across the entire ranch area. Furthermore, high-sensitivity LoRa modules are employed to enhance signal reception effectiveness.

3.2 Challenges and Solutions

(1) Insufficient Signal Coverage: In densely wooded or complex terrain areas, signal coverage may be inadequate. To address this, we employ additional relay nodes or adjust the gateway locations to expand signal reach.
(2) Environmental Interference: Meteorological conditions (such as lightning, rainfall) and electromagnetic interference can compromise signal quality. To mitigate these issues, we utilize LoRa modules with robust anti-interference capabilities and incorporate software algorithms for signal filtering and correction.
(3) Equipment Costs and Maintenance: Device costs and maintenance represent crucial considerations during implementation. To optimize costs, we select cost-effective LoRa modules and devise a system architecture that facilitates easy maintenance and upgrades.

4 System Communication Parameters

Spread Spectrum Communication is a digital signal communication technology that expands signal bandwidth multiple times for communication. Its basic feature is that the bandwidth of the transmitted information signal is much greater than the bandwidth of the information itself. Spread spectrum technology uses a spread spectrum function that is independent of the transmitted information to modulate it. At the receiving end, the same spread spectrum function is used to demodulate the spread spectrum signal and restore the transmitted information.

The communication principle provides the definition of Shannon's theorem: the maximum error free information transmission rate that can be achieved for a given channel bandwidth in the presence of noise. The formula is shown in Eq. 1.

$$C = B \times \log_2\left(1 + \frac{S}{N}\right) \tag{1}$$

In the formula, C is the channel capacity, B is the channel bandwidth, S is the average signal power, and N is the average noise power. When the transmission rate C of the signal is constant, increasing the signal bandwidth can reduce the requirement for signal-to-noise ratio. Theory suggests that the anti-interference performance of various spread spectrum systems is related to the bandwidth ratio of the spread spectrum signal after information spectrum expansion. The ratio of the spread spectrum signal bandwidth B to the information bandwidth ΔF is generally referred to as the processing gain G_P. The formula is shown in Eq. 2.

$$G_p = \frac{B}{\Delta F} \tag{2}$$

The processing gain reflects the degree of improvement in the signal-to-noise ratio of the spread spectrum communication system, and it is an important performance indicator of the spread spectrum system.

Definition of anti-interference tolerance M_J for spread spectrum communication systems, as shown in Eq. 3

$$M_J = G_P - \left[\left(\frac{S}{N}\right)_o + L_S\right] \tag{3}$$

In spread spectrum communication systems, the anti-interference tolerance M_J is related to the processing gain G_P. By increasing the processing gain through spread spectrum, the anti-interference tolerance is also increased.

In LoRa technology, spread spectrum technology uses multiple information chips to represent each bit of effective information. The transmission rate of information is called symbol rate (*Rs*), and the ratio of the spread spectrum chip rate to Rs is the spread spectrum factor (*SF*).Indicates the number of symbols sent for each information bit. The formula is shown in Eq. 4.

$$SF = SymbolRate/ChipRate \tag{4}$$

The spread spectrum technology used in LoRa technology is based on direct sequence spread spectrum. The original signal of user data is XOR operated with pseudonoise code, that is, during spreading, each bit of data is multiplied by the spreading factor to generate a transmission signal stream. At the same transmission rate, the actual amount of data transmitted is reduced. Therefore, the larger the spreading factor, the lower the bit rate.

In communication systems, the wireless transmission distance is determined by the sensitivity of the receiver and the output power of the transmitter, and the difference between the two is called link budget. The output power is limited by standard specifications, so it is necessary to increase communication distance by increasing sensitivity. The sensitivity formula is shown in Eq. 5.

$$\text{Sensitivity} = -174 + NF + 10\lg B + 10\lg SNR(\text{dB}) \tag{5}$$

As can be seen from the above equation, the receiver sensitivity is the sum of the dB values of noise coefficient (*NF*), signal bandwidth (*B*), and demodulation signal-to-noise ratio (*SNR*), minus 174 dB.

In summary, LoRa utilizes direct sequence based spread spectrum technology, which works by expanding the transmission signal bandwidth, improving signal-to-noise ratio, and increasing the spread spectrum processing gain to enhance the anti-interference tolerance of the communication system. Thus, under power limited conditions, the sensitivity of the receiver is improved, resulting in a strong link budget and longer transmission distance for the system. Therefore, LoRa spread spectrum technology changes the balance between transmission power consumption and transmission distance to achieve low power consumption and long-distance performance characteristics.

5 System Configuration and Data Transmission

The system configuration in this study takes into account the unique characteristics of the pasture environment meticulously. For instance, the adjustment of air data rates and transmission power has been conducted to cater to varying distances and signal quality requirements. Furthermore, the deployment of multiple communication channels is employed to enhance the reliability and stability of the system. In the realm of data processing, we attach paramount importance to both the accuracy and timeliness of data. In response to potential data anomalies or losses inherent in the pasture environment, we have devised a robust mechanism of data verification and error correction, thereby ensuring the integrity and reliability of the data. This approach not only aligns with academic rigor but also underscores the practicality and robustness of the system in real-world applications.

The data packet format of LoRa signal is shown in Fig. 1, which includes two different formats of wireless data packets: explicit and implicit. Explicit packets include n-bit preamble symbols, n-bit header symbols, payload data, and cyclic checksum of the payload, while implicit packets include n-bit preamble symbols, payload data, and cyclic checksum of the payload, without a header portion. The header mainly includes information such as load data length, encoding rate, and whether there is CRC verification.

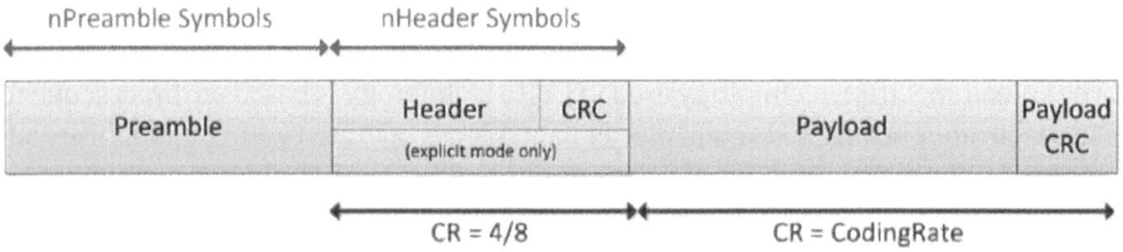

Fig. 1. LoRa packet

The transmission time of LoRa packets is the sum of the n-bit preamble symbol transmission time and the load data transmission time. The formula is shown in Eq. 6.

$$T_{packet} = T_{preamble} + T_{payload} \tag{6}$$

Among them, $T_{preamble}$ is the transmission time of the preamble, and its formula is shown in Eq. 7.

$$T_{preamble} = (n_{preamble} + 4.25) \times T_{sym} \tag{7}$$

In Eq. 7, $n_{preamble}$ is the number of preamble symbols, T_{sym} is the symbol period, which is the reciprocal of the symbol rate. The formula is shown in Eq. 8.

$$T_{sym} = \frac{1}{R_s} \tag{8}$$

Secondly, $T_{payload}$ is the transmission time for transmitting load data, and its formula is shown in Eq. 9.

$$T_{payload} = n_{payload} \times T_{sym} \tag{9}$$

where $n_{payloade}$ is the number of symbols for the payload.

The intelligent ranch environmental monitoring system studied in this paper consists of three different functional LoRa sensor nodes, LoRa gateways, and upper computers. Three types of sensors collect environmental temperature and humidity data, light intensity data, and PM2.5 concentration data respectively. The data is packaged through LoRa modules and transmitted remotely to the LoRa gateway with LoRa protocol transparent transmission. After receiving the data, the gateway integrates the three data points and transmits them to the PC for data display through serial communication. The system adopts a three-point to one information transmission mode, so the LoRa module adopts a transparent transmission mode, with addresses set to 0XFFFF, so that the module is in transparent transmission and broadcast listening mode; The configuration adopts AT instructions and is directly implemented through the UART program; Communication channel setting 23, air speed 19.2 kbps, transmission power 20 dBm, transmission status set to transparent transmission, serial port configuration USART3, baud rate 115200 bps. After receiving the sensor initialization command, the node will perform GPIO initialization and clock initialization in the sensor initialization function. After successful detection, data will be read.

The data format for temperature and humidity node collection is 5 bytes, where the first 4 bytes of data, namely temperature and humidity data, are taken for data encapsulation; When collecting data from photosensitive sensors and PM2.5 sensors, ADC conversion is performed. Therefore, ADC initialization, clock configuration, and channel configuration need to be performed in the sensor initialization function. The output data is all float type 4 bytes, and after reading, the data is encapsulated. After receiving the initialization instruction of the lora_receive function, the gateway receives data and reads the starting symbol to read the next byte node address. After reading the four bytes of data for different received node addresses, the gateway performs different forms of data type transformation. After parsing, the data is transmitted through the serial port to the PC serial port debugging assistant for unified display.

6 Conclusion

LoRa technology exhibits remarkable autonomous development capabilities within the smart ranch environmental monitoring system, providing robust support for the aggregation of multi-type information in livestock Internet of Things (IoT) systems. This paper delves into the independent application of LoRa technology within LPWAN for multi-point-to-single-point communication in smart ranch environmental monitoring systems, showcasing the exceptional autonomy of LoRa modules in IoT technology development. The study validates the feasibility and effectiveness of LoRa technology in complex ranch environments. Notably, the methodologies for aggregating information from diverse nodes during multi-point-to-single-point communication serve as a cornerstone for future LoRa technology implementations in livestock IoT systems where users autonomously develop systems for multi-type information aggregation. As such, we are committed to continually optimizing system designs and technical solutions to address emerging challenges and further propel the intelligent and informatized transformation of the livestock industry.

References

1. Kannayeram, G., Madhumitha, M., Mahalakshmi, S., et al.: Smart environmental monitoring using LoraWAN. In: Sikdar, B., Prasad Maity, S., Samanta, J., Roy, A. (eds.) Proceedings of the 3rd International Conference on Communication, Devices and Computing. ICCDC 2021. LNEE, vol. 851, pp. 513–520. Springer, Singapore. (2022). https://doi.org/10.1007/978-981-16-9154-6_48
2. Tang, X., Liu, S., Zang, C.: Design of tobacco storage environment monitoring system based on Lora technology. In: ITM Web of Conferences, 2022. https://doi.org/10.1051/itmconf/20224501064
3. Ou, Y., Ma, C., Hao, Y.: Design of environmental monitoring and control system for greenhouse based on long range wireless technology. In: Proceedings of the 2021 8th International Conference on Wireless Communication and Sensor Networks, 2021. https://doi.org/10.1145/3461717.3461721
4. Jiang, J., Zhu, W., Xie, X., et al.: Indoor environment quality monitoring and evaluation system based on LoRa communication. Comput. Commun. (004), 010 (2022)
5. Purnomo, F.A., Yoeseph, N.M., Yulianto, A., et al.: Development of wind monitoring systems with LoRA technology. In: IOP Conference Series: Earth and Environmental Science, p. 989, 2022. https://doi.org/10.1088/1755-1315/989/1/012011
6. Promput, S., Maithomklang, S., Panya-Isara, C.: Design and analysis performance of IoT-based water quality monitoring system using LoRa technology. TEM J. (2023). https://doi.org/10.18421/tem121-04
7. Huiying, C., Fangzhen, L., Peng, L., et al.: Intelligent aquaculture environment monitoring system based on Lora communication technology. INMATEH-Agric. Eng. 109–120 (2021). https://doi.org/10.35633/INMATEH-63-11
8. Krianovi, V., Grgi, K., Spisic, J., et al.: An advanced energy-efficient environmental monitoring in precision agriculture using LoRa-based wireless sensor networks. Sensors (Basel, Switzerland) **23** (2023). https://doi.org/10.3390/s23146332
9. Kumar, R.P., Sushmanth, C., Girish, K.V.: LoRa based garbage monitoring system over IoT. Int. Organ. Sci. Res. **12**(5), 33–44 (2022)

Simulation and Analysis of Harmonic Reducer Dynamics Based on Adams

Fan Zhenduo(✉)

Heilongjiang Polytechnic, Harbin 150028, Heilongjiang, China
fanzhenduo@126.com

Abstract. The harmonic reducer is a new type of reducer, developed on the theoretical basis of harmonic gearing technology, the harmonic reducer was first used in the field of aerospace technology. Simulation and analysis of harmonic reducer dynamics is an essential direction in the field of harmonic reducer research; through the use of three-dimensional modelling software SolidWorks and mechanical system dynamics simulation software ADAMS, the establishment of harmonic reducer transmission parts the three-dimensional solid simulation model, discusses the model establishment method and the simulation process, the study of the flexible wheel and the rigid wheel change force rule, to explore in the flexible wheel and the rigid wheel mesh stiffness and the mesh damping of a specific condition, the wave generator length of the long axis and the short axis of the length of the change rule of the impact of the mesh force change rule, to provide a basis for the harmonic speed reducer research.

Keywords: Harmonic Reducer · Dynamic Analysis · Meshing Force

1 Preamble

According to Dr. Musser C.W. instructions "Harmonic Drive relies on flexible parts that generate elastic mechanical waves to transmit motion and force" [1]. Because of its small size, lightweight, large transmission ratio, transmission precision, and other advantages, it has been widely used in robotics, precision transmission, and other fields. Harmonic gear transmission in the transmission field has been rapidly developing; scholars at home and abroad on the harmonic gear transmission research are increasingly in-depth compared with the level of foreign harmonic transmission technology. According to Ruifeng W instructions "There are still large gaps in our country, the production of domestic harmonic transmission device there are still limited customization, small-scale production, manufacturing costs and high limitations" [2]. The dynamics simulation and analysis of harmonic reducers are of great significance for the study of harmonic reducer motion characteristics and the improvement of harmonic reducer dynamics performance. This paper will be explore the rule of change of harmonic reducer meshing force through kinetic simulation analysis to provide a theoretical basis for further research on harmonic reducer.

X. Li et al. (Eds.): 6GN 2024, LNICST 615, pp. 143–152, 2025.
https://doi.org/10.1007/978-3-031-93193-2_12

2 Harmonic Reducer Drive Principle

The Harmonic Reducer, also known as Harmonic Gear Reducer, is a high-precision, high-efficiency speed reduction device that is primarily used to reduce speed and increase torque. Harmonic reducer is mainly composed of a rigid wheel, a flexible wheel and a wave generator. There are three essential components, as shown in Fig. 1: the wheel teeth of the rigid wheel and flexible wheel are mostly involute profiles. Harmonic transmission mainly has double-wave transmission and three-wave transmission two methods, double-wave transmission flexure stress is smaller, simple structure, can get a larger transmission ratio, is currently the most widely used One of the most widely used. According to Zhenlin X instructions "the rigid wheel and flexible wheel have the same circumference, however number of teeth is not equal, usually using a rigid wheel, and the number of teeth on the flexible wheel is equal to the difference between the number of waves" [3]. The elliptical wave generator is mounted inside the flex wheel. The length of its long axis is a, and the length of its short axis is b, as shown in Fig. 2.

The wave generator's long axis size is slightly larger than the inner diameter of the flexible wheel when the wave generator is loaded into the flexible wheel, forcing the flexible wheel to form an elliptical shape so that the wave generator in the wave generator at the two ends of the long axis of the flexible wheel gear teeth and the rigid wheel gear teeth are in a fully meshed state, the short axis of the two ends of the flexible wheel teeth and the rigid wheel gear teeth are in the state of disengagement, and the other segments of the flexible wheel teeth in the transition state. According to Shijie C instructions "During the transmission process, the teeth of the flexible wheel and the rigid wheel are constantly engaged, engaged, engaged and disengaged in the four sequential cycles of state to realize the transfer of motion and power between the flexible wheel and the rigid wheel" [4].

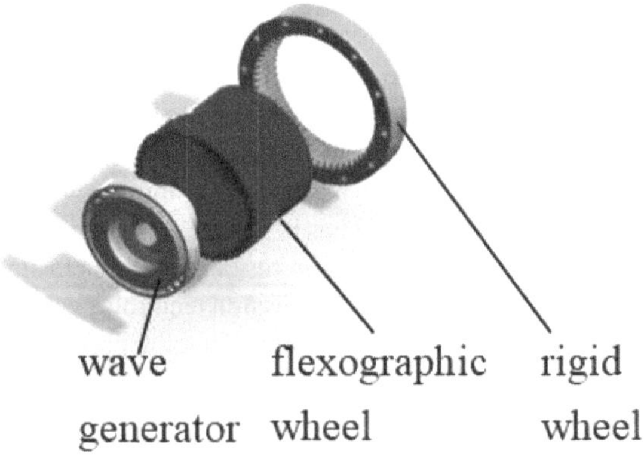

Fig. 1. Main components of the harmonic reducer

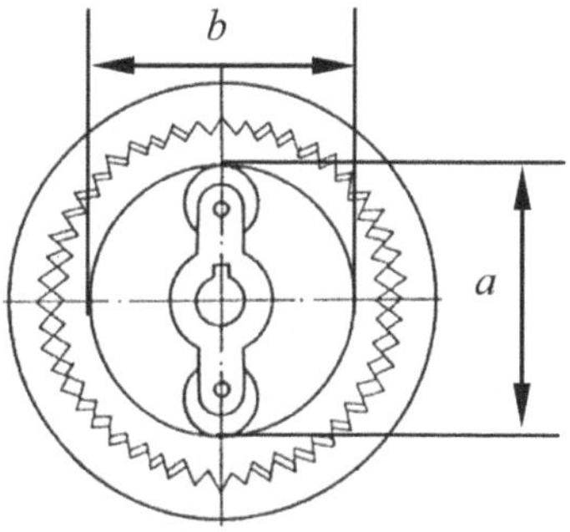

Fig. 2. Harmonic Reducer Drive Principle

When the elliptical wave generator is mounted inside the flexure wheel, the centerline of the flexure ring changes from a circle with radius r to an ellipse. Since there is a flexible bearing between the inner bore of the flex wheel and the cam of the wave generator, According to Yibo D instructions "the cam profile can be considered as an inner isometric curve of the elliptical wave generator"[5].

Long half-axis of the ellipse of the centerline of the flexor wheel a

$$a = r + mw_o \quad (1)$$

included among these the w_0 which is means radial deformation coefficient of flexible wheel

short half-axis of the ellipse in the center of the flexor wheel b

$$b = 1/9[(12r - 7a + 4a(3R - 2a))] \quad (2)$$

If the perimeter of the convex contour line is taken approximately as $\pi(a + b)$, then the above equation can be changed to:

$$b = 2R - a = R - w_o m \quad (3)$$

In practice, the difference between the values obtained from the above equation is less than one ten-thousandth of the value of b; therefore, the design of the calculation according to the above formula will be able to meet the accuracy requirements.

During the harmonic gearing process, one of the three main components of the rigid wheel, the flexible wheel, and the wave generator can be fixed, and one of the other two components serves as the master and the other as the follower.

In practice, the rigid wheel is fixed in most cases, with the wave generator as the active part and the flexible wheel as the driven part. According to Fig. 3; Let the number of teeth of the rigid wheel be z_1 and the number of the flexible wheel be z_2; then, its transmission ratio is:

$$i_1 = z_1/(z_2 - z_1) \quad (4)$$

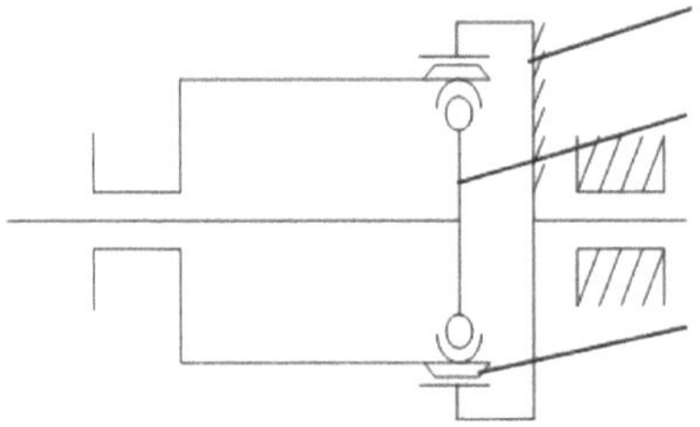

Fig. 3. Rigid Wheel Fixed Transmission Sketch

3 Simulation of Transmission Dynamics of Harmonic Reducer

3.1 Theoretical Analysis of Gear Meshing Dynamics

Compared with ordinary gear meshing, harmonic gear meshing is more complex; when the wave generator is installed inside the flexible wheel, the shape of the flexible wheel becomes elliptical with the wave generator. In the transmission process, the flexible wheel gear teeth and the rigid wheel gear teeth are constantly in the meshing, meshing, meshing and disengagement of the four sequential cycles of the state so that the flexible wheel gear teeth and the rigid wheel gear teeth meshing force is in the process of constant change. Dynamic simulation and analysis of harmonic reducer using ADAMS can observe the different situations of the change of the engagement force curve and find the weak points in the process of change, which is of great significance for the study of the engagement force.

According to Bin L instructions "Due to the complexity of the harmonic transmission, the meshing force is solved based on the nonlinear dynamics equations of the gear system" [6], which are expressed as follows:

$$\mathrm{Mx} + \mathrm{Cx} + \mathrm{K(t)}\{\mathrm{X} + \mathrm{X_s} + \mathrm{e(t)}\} = \mathrm{P_s} \quad (5)$$

In this expressed the M which is means the system quality matrix, the C which is means the system damping matrix, the K(t) which is means the system stiffness matrix, the X which is means dynamic displacement vectors, the X_S which is means the static relative displacement vector, the e(t) which is means gear synthesis error, the P_S which is means static loads. And the K(t) includes the stiffness of the gear train, the stiffness of the structural system and the bond stiffness, but also the time-varying integrated stiffness of the gear mesh.

3.2 Motion Simulation

The harmonic reducer studied and analyzed in this paper is a single-stage harmonic reducer. Firstly, the harmonic reducer components are modelled by the three-dimensional modelling software Solidworks. The model contains all the necessary parts, such as wave generators, flex wheels, rigid gears, etc., and ensures that the relative positions and dimensions between the parts are accurate. Exporting a 3D model to a format supported by Adams software ensures that the geometric features and assembly relationships of the model are preserved during the export process. Import the model in Adams software and make the necessary checks and adjustments to ensure that the model is correct and complete in Adams.

To facilitate the study of the system, the following assumptions are made for the kinetic model:

1. Neglecting manufacturing errors, the assembly factor is zero.
2. Each rotation constraint is an ideal constraint, ignoring the rolling between the bearing and the shaft.
3. The effect of gear deformation on the meshing force is not considered.

The model is tested before starting the simulation to eliminate errors implicit in the modelling process and ensure the simulation can proceed smoothly. The above assumptions allow the theoretical analysis to exclude the interference of secondary factors, making the model more concise and easy to handle, eliminating model uncertainty due to inaccuracies in the manufacturing process and bias in the assembly process, reducing computational complexity, and focusing on the core kinetic characterization.

In Adams, the necessary kinematic vice and contact forces are defined according to the actual working principle of the harmonic reduce, the rotating vice is used to simulate the relative rotation between the wave generator and the flexure wheel, while the fixed vice is used to fix the components such as the rigid gear.

1. Applying motion subs

   Apply fixed vice to the rigid wheel and the earth, and add rotating vice between the wave generator, the flexible wheel and the earth. When adding the moving vice, pay attention to the selection order and relative position of the components to ensure that the direction and position of the moving vice are correct.
2. Setting the Motion Subparameters

   For the applied fixed vice and rotary vice, the initial position, rotation range and other corresponding parameters are set.
3. Setting Contact Properties

   Define the contact pairs between the wave generator and the flex wheel, and between the flex wheel and the rigid gear, and set the parameters of stiffness, dampingand and force exponent according to the material properties of the harmonic reducer and the working requirements.
4. Contact Force

   After defining the contact pairs and setting the contact properties, adams automatically calculates the contact forces and applies them to the corresponding parts.

According to Guoqi X instructions "In ADAMS, there are two main methods for contact force calculation: one is the compensation method, and the other is the impact function method" [7]. The compensation method is based on the concept of the coefficient of recovery in physics, and the energy loss during contact is modeled by setting the penalty and compensation coefficients, which is more intuitive for understanding the physical nature of the contact force. When applying the compensation method, it is necessary to determine the penalty coefficient and the compensation coefficient, which are used less often due to the complexity of determining the penalty coefficient and the fact that the coefficient does not respond to the actual two-contact-body forces by either too large or too small a change in the coefficient; According to Wei W instructions "Impact function method does not need to determine such a coefficient, a more realistic response to the force situation, so for the impact function method is more often used" [8]. The impact function method is able to more accurately simulate the physics of the contact process by using a detailed mechanical model to calculate the contact forces, including elastic forces due to mutual incision and damping forces due to relative velocities. The function is defined as:

$$Kx^{e}+d\{F_s(x, 0, 0, d_e, C_{max})\}/dX=f \qquad x<0$$

$$f=0 \qquad x\geqslant 0$$

In this expressed the K which is means Rigidity coefficient, the x which is means the distance variable between two objects, the e which is means nonlinear index, the F_s which is means stepfunction, the C_{max} which is means damping factor, the d_e which is means depth of breakdown at maximum damping.

When $x \geq 0$, The two objects are not in contact, and the magnitude of the contact force is zero. When $x < 0$, Two objects come into contact, and the magnitude of the contact force is related to the stiffness coefficient, deformation, nonlinear exponent, damping coefficient and depth of breakdown.

3.3 Analysis of the Simulation Results of the Meshing Force of Harmonic Drive Components

In order to simulate the actual working condition in the simulation process, a rotational speed drive is applied to the input axis using the STEP function, which is set to STEP (0,0d,0,10000d), and the simulation time is selected to be 2s, with a step size of 200. The method of this paper to study the factors affecting the engagement force is to change the length of the wave generator elliptical cam long and short axes and observe the change of the engagement force curve so as to explore the change rule of the engagement force. In order to cooperate with their own four degrees of freedom parallel robot research work, this paper selected a single-stage harmonic reducer wave generator oval cam long axis a for 103 mm, short axis b for 93 mm. In this paper, the effect of the change in the length of the long shaft on the change in the engagement force curve for a certain length of the short shaft and the effect of the change in the length of the short shaft on the change in the engagement force curve for a certain length of the long shaft are investigated respectively. The following figures show the effect of change in the length of the long axis on the engagement force for the case of short axis b = 93 mm (as shown in Fig. 4) and the effect of change in the length of the short axis on the engagement force for the case of long axis a = 103 mm (as shown in Fig. 5).

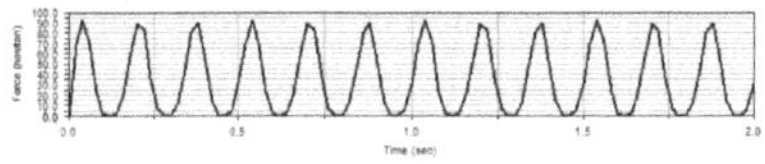

(a) a=95mm, b=93mm

Peak engagement force (proximity) 90N

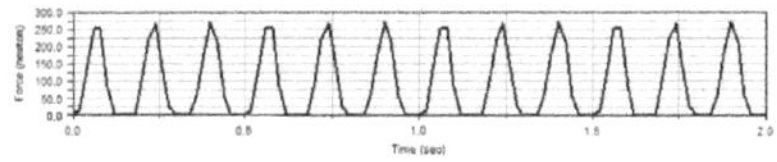

(b) a=98mm, b=93mm

Peak engagement force (proximity)280N

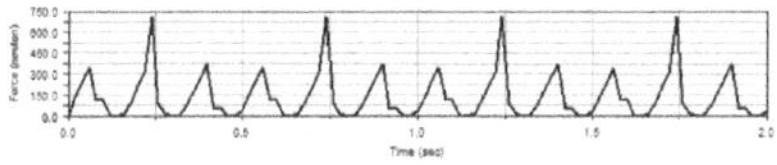

(c) a=101mm, b=93mm

Peak engagement force (proximity) 730N

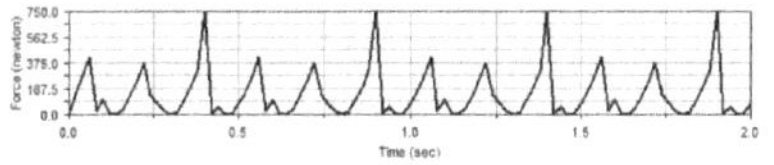

(d) a=102mm, b=93mm

Peak engagement force (proximity) 750N

Fig. 4. Short axis b = 93 mm, long axis variation

Based on Fig. 4, the data was collected and organized (as shown in Table 1) and it can be found that, when the short axis length is certain, the peak engagement force decreases from 750N attachment to 90N attachment as the long axis value continues to decrease.

Based on Fig. 5, the data was collected and organized (as shown in Table 2) and it can be found that, when the length of the long axis is certain, the peak engagement force decreases from near 730 N to near 350 N as the value of the short axis b gradually increases.

The simulation results all show that, the peak meshing force decreases as the length of the long axis gets closer to the length of the short axis.

Using the same method, this paper also investigates the effect of the change in the length of the long axis of the wave generator on the engagement force in the case of the short axis b = 93 mm, 94 mm, 95 mm, 96 mm, 97 mm and the effect of the change in the length of the short axis on the engagement force in the case of the a = 103 mm, 102 mm, 101 mm, 100 mm, 99 mm, respectively, which are organized into the curves shown in the following figures:

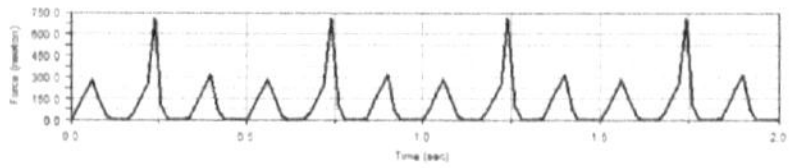

(e) a=103mm, b=86mm

Peak engagement force (proximity)730N

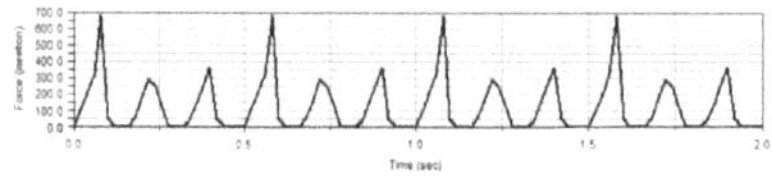

(f) a=103mm, b=89mm

Peak engagement force (proximity)700N

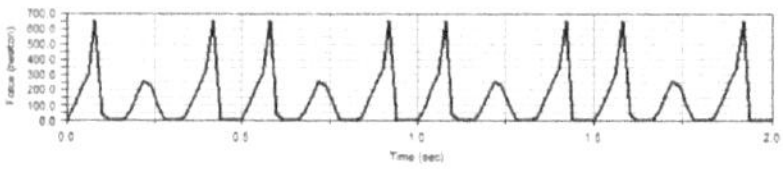

(g) a=103mm, b=92mm

Peak engagement force (proximity)680N

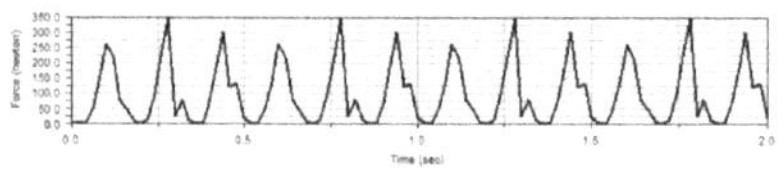

(h) a=103mm, b=93mm

Peak engagement force (proximity)350N

Fig. 5. Short axis a = 103 mm, short axis variation

Table 1. Short axis length fixed, long axis variable

	a = 102, b = 93	a = 101, b = 93	a = 98, b = 93	a = 95, b = 93
Peak engagement force (proximity)	750N	730N	280N	90N

Table 2. Long axis length fixed, short axis variable

	b = 93, a = 103	b = 92, a = 103	b = 89, a = 103	b = 86, a = 103
Peak engagement force (proximity)	730N	700N	680N	350N

The curves shown in Fig. 6 represent the effect of the change in the length of the long axis on the engagement force in the case of short axis b = 93 mm, 94 mm, 95 mm, 96 mm, 97 mm. The figure shows that when the length of the long axis is closer to the length of the short axis, i.e., the length of the transverse coordinate L is getting smaller and smaller, the engagement force curve also becomes smoother, and the transmission is more stable. Furthermore, it can be seen that among all the curves, the curve with b =

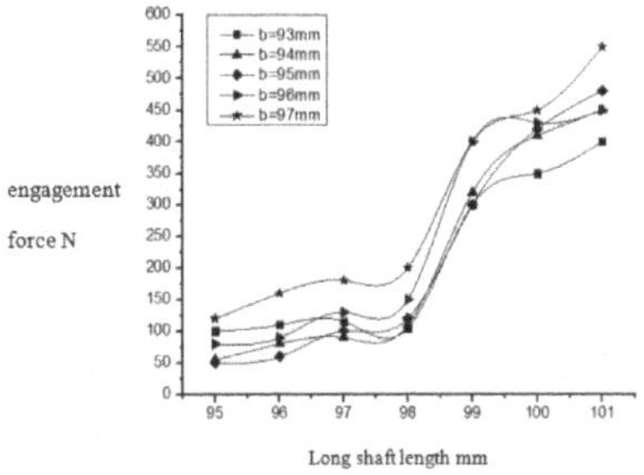

Fig. 6. Change in the length of the long axis

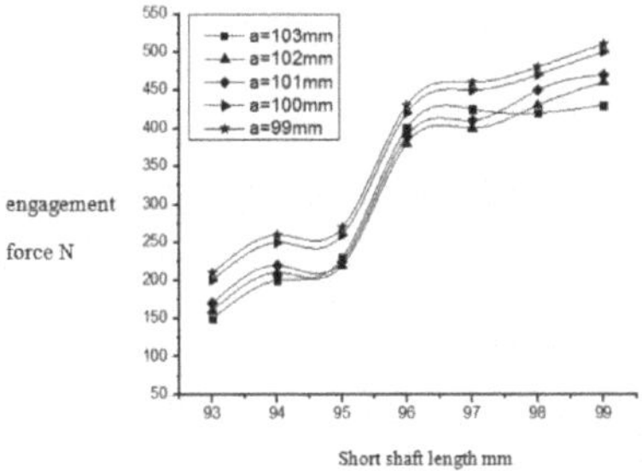

Fig. 7. Change in short-axis length

93 mm is smoother, which indicates that for the wave generator elliptical cam, the short shaft length is more appropriate when it is 93 mm. The curves shown in Fig. 7 represent the effect of the change of the short axis length on the engagement force in the case of the long axis a = 103 mm, 102 mm, 101 mm, 100 mm, 99 mm; it can be seen from the figure that when the short axis length becomes longer and longer, i.e., when the short axis length is closer to the change of the long axis, the curve of the engagement force becomes smoother. The transmission is smoother, and it can be seen that among all the curves, the long axis a = 103 mm, the curve changes more smoothly when the short axis length gets closer to the long axis length, indicating that for the wave generator elliptic cam, the long axis length of 103 mm is more suitable for harmonic reducer motion. Based on the above data, it can be known that the engagement force curve is smoothest and the transmission is smoother when the long axis a = 103 and the short axis b = 93.

By analyzing the simulation results of the harmonic reducer transmission parts, it can be seen that the change in the length of the long and short shafts of the elliptical cam of the harmonic reducer wave generator is an essential factor that affects the meshing force between the flexible wheel and the rigid wheel. The smoothness of the harmonic transmission can be effectively improved if the length of the long and short shafts is changed reasonably. Therefore, if the design stage of the harmonic reducer can find the appropriate ratio of the long and short axis, it is of great significance to improve the smoothness of the harmonic reducer.

4 Conclusion

In this paper, through the harmonic reducer dynamics simulation analysis research, explored the harmonic reducer flexible wheel and rigid wheel meshing force change law, understand that in other conditions must be the case, the wave generator elliptic cam long and short axis length change is a significant factor affecting the flexible wheel and rigid wheel meshing force. This for the future study of four degrees of freedom parallel robot in the process of the use of harmonic reducer has a more in-depth understanding of the simulation and analysis of the study can be used in the future according to the actual situation as far as possible to avoid the emergence of the situation of the sudden change of the meshing force so that the parallel robot movement will be additional stable.

References

1. Musser, C.M.: The Harmonic Drive Break through in Mechanical Drive Design. Machine Design, 1960
2. Ruifeng, W.: Harmonic Gearing Overview, 2019
3. Zhenlin, X.: Harmonic Reducer Research Status and Problems, 2020
4. Shijie, C.: Research on anti-fatigue characteristics and parameter optimization of harmonic reducer flexure wheel, 2021
5. Yibo, D.: Mechanical Characterization and Structural Optimization of Harmonic Reducer Flex Wheel, 2019
6. Bin, L.: Fatigue life of flexible bearing of harmonic reducer and its mechanical characterization, 2023
7. Guoqi, X.: Finite Element Analysis Study of Harmonic Gear Drive Flex Wheel, 2005
8. Wei, W.: Simulation and analysis of six-axis handling robot dynamics based on ADAMS, 2024

V2V Communication Systems for Advanced Truck Platooning at Freight Hubs

Tao Song[1], Jinlong Hao[1(✉)], Luxi Cheng[1], Hui Li[1], Xiuhua Li[1(✉)], and Penghua Li[2]

[1] School of Big Data and Software Engineering, Chongqing University, Chongqing, China
haojl@changan.com.cn, hli@cqu.edu.cn, lixiuhua1988@gmail.com

[2] School of Big Data and Software Engineering, Chongqing University of Posts and Telecommunications, Chongqing, China
liph@cqupt.edu.cn

Abstract. Truck platooning can improve traffic efficiency while reducing fuel consumption by using vehicle-to-vehicle (V2V) communication based on 6G communication. This paper adopts a CACC model to control the platoon trucks in freight hubs for maximizing the overall profit of a short-medium distance logistic transportation system. It evaluates various dispatch strategies, accounting for environmental benefits and extra costs such as waiting times and goods delays in platoon formation. Utilizing real heavy-duty truck data, the study computes each truck's VSP distribution to determine emission and fuel consumption rates. We introduce an enhanced PSO algorithm tailored for mining truck platoons. Using an advanced Nguyen and Dupius transport network model, different departure intervals and maximum platoon sizes are assessed. Results indicate that balancing ecological gains with additional wait costs significantly impacts profitability. The optimal strategy found involves departures every 30 min with a platoon limit of 5 trucks, leading to a peak profit of 512.70$ using the proposed algorithm.

Keywords: Truck platooning · freight hubs · V2V communication · CACC model · ecological benefit

1 Introduction

The transportation sector is universally acknowledged for its substantial impact on environmental deterioration and the strain on energy resources. In particular, heavy-duty trucks are known to guzzle significant quantities of diesel fuel, resulting in the release of various pollutants (e.g., carbon monoxide(CO), carbon dioxide(CO2), etc.)into the atmosphere, which poses a threat to the ecological balance [1,2]. According to projections, by 2030, the diesel consumption by trucks in China is expected to soar to 221 million tons.

X. Li et al. (Eds.): 6GN 2024, LNICST 615, pp. 153–166, 2025.
https://doi.org/10.1007/978-3-031-93193-2_13

With the in-depth study of 6G networks by researchers, it is found that 6G networks have higher data transmission rates, a wider coverage area, and are better suited for Vehicle-to-Vehicle (V2V) communication than 5G networks. Currently, the latency of 5G networks is usually around 1 millisecond, but the latency of 6G networks will be reduced to the microsecond level. Relevant research shows lower V2V communication latency means that vehicles have more timely reaction time in the process of high-speed vehicle driving [3]. The vehicle-to-vehicle (V2V) [4,5] technology can realize the communication of vehicle to vehicle and vehicle to environment, enabling vehicles to obtain surrounding information in real time. Regarding truck platooning, V2V communication enables trucks to follow each other closely, thus reducing air drag and enhancing fuel economy [6]. It is highly important for the development of an eco-friendly transportation system.

Lately, truck platooning has been attracting more and more attention. Some literature is dedicated to assessing the performance of truck platooning in terms of fuel savings, operational costs, or equipment utilization through dispatching optimization algorithms [7,8]. Based on the fuel consumption of the platoon, Chen *et al.* put forward a mixed-integer second-order-cone programming model to minimize the total operation cost and solve the scheduling problem for container transshipment between two seaport terminals [9]. Xu *et al.* formulated a problem to route the trucks to destinations on time using the least amount of fuel while satisfying the drivers' mandatory breaks [10]. The authors performed a co-optimization model on planning speed and splitting power to reduce the cost of hybrid following trucks. The predictive model is used to implement the optimization strategy [11]. These studies usually pay attention to the algorithm theories or optimization models for energy efficiency or the costs of platoon trucks. However, the ecological effect of the platoons is another crucial factor that ought to be meticulously taken into account.

In this paper, we put forward an ecological truck platoon optimization approach to achieve the maximum overall profit of medium and short-distance logistics transportation systems. Taking into account ecological benefits and additional costs, we put forward an improved particle swarm optimization (PSO) algorithm for mine truck platoon, aiming to reduce fuel consumption at freight hubs. We use the actual operational data of heavy-duty trucks between cities as the working conditions for the first truck in the platoon. By adopting the cooperative adaptive cruise control (CACC) model, we obtain the operating conditions for the following trucks in the platoon. In addition, we compute the VSP distribution within various speed ranges and obtain the emission and fuel consumption factors for each truck in the platoon. The experimental results indicate that the proposed algorithm is appropriate for planning truck platoons in freight hubs.

2 Ecological Characteristics of CACC Truck Platoon

In CACC vehicle model, each vehicle is viewed as a real-time signal source. It is capable of obtaining information regarding surrounding vehicles and the environment, and transmitting its current conditions (e.g., position, speed, and acceleration, etc.) to other vehicles based on V2V communication (as shown in Fig. 1). This seamless sharing of information among the vehicles enables coordinated movements and improves overall driving efficiency and safety [12,13].

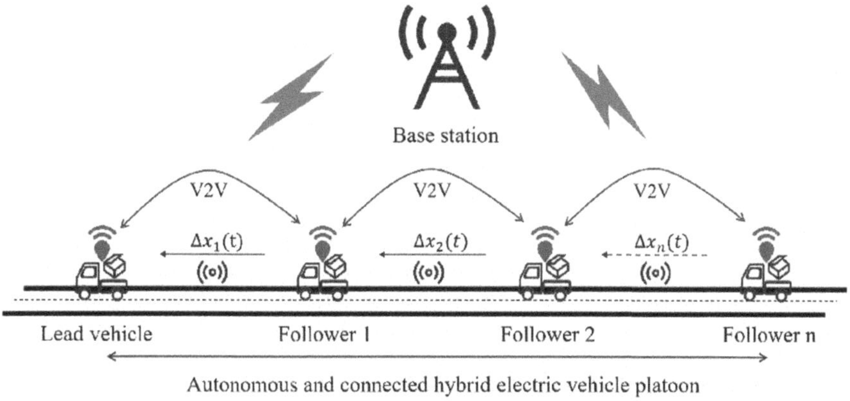

Fig. 1. Illustration of V2V communication for truck platooning

The VSP refers to the instantaneous power per unit of mass of the vehicle(kw/ton), which represents a crucial parameter in vehicle dynamics and performance analysis [14]. The VSP of a truck can be calculated by

$$VSP_t = \frac{Bmgv_t + \frac{1}{2}\rho ACv_t^2 + mv_t a_t}{f_{scale}}, \tag{1}$$

where v_t is the truck speed(m/s) at time t; a_t is the truck acceleration(m/s^2); m is the total truck mass (ton); g is the acceleration of gravity (9.8 m/s^2); B is the rolling resistance factor which depends on the road surface, tire type and pressure; A is the windward area when the truck is driving (m^2); ρ is the air density (kg/m^3); C is the aerodynamic resistance factor which is determined by the temperature and pressure conditions of air; f_{scale} is the scale factor.

When the trucks operate in a platoon formation, the aerodynamic drag experienced by the rear trucks will be reduced [15]. As the alteration of the air resistance coefficient of a truck primarily depends on the distance to the truck in front [16]. Hence, the aerodynamic resistance factor can be modified as

$$C = C_{D0}(1 - \frac{C_{D1}}{C_{D2} + s_n}), \tag{2}$$

where C_{D0}, C_{D1} and C_{D2} are the empirical coefficients; s_n ($n \geq 1$) is the distance between n-th truck and its front truck (m). When $n = 1$, we believe that $s_n \rightarrow \infty$ and $C = C_{D0}$ at the same time, which means the aerodynamic resistance factor of the lead truck is C_{D0}.

The influence of VSP on fuel consumption is more significant than that of other commonly utilized parameters, like speed and acceleration [17]. Therefore, the ecological characteristics along with the operating conditions of platoon trucks can be calculated as follows.

Fuel consumption factor: the fuel consumption factor of the truck within different speed ranges can be computed as

$$\begin{aligned} CF_k &= \sum_j cr_j \cdot D_{j,k}) \cdot \frac{3600}{v_k} \\ &= \sum_j (\frac{ct_j}{t_j} \cdot D_{j,k}) \cdot \frac{3600}{v_k}, \end{aligned} \tag{3}$$

where CF_k is the consumption factor in k-th speed range(g/km); cr_j, ct_j and t_j is the consumption rate(g/s), total fuel consumption(g) and driving time(s) in j-th VSP bin, respectively; v_k is the mid-speed in k-th speed range(km/h).

3 Problem Formulation and Algorithm Design

In freight hubs, the trucks that arrive earlier need to wait for the ones that arrive later to form a platoon, thus resulting in additional waiting costs (for instance, the idle cost for trucks and drivers) and the delayed penalty cost of goods. The operating mode of freight hubs is shown in Fig. 2. Suppose the time interval is T, and nine trucks reach the freight hub in the same duration. Among the nine trucks, there are three square-marked trucks, three circular-marked trucks, and three triangular-marked trucks, representing various driving routes. The scheduling of these trucks can enable them to leave the freight hub in platoons for maximizing the total system benefits. While considering the balance of the ecological benefits and additional costs for forming the platoons, the nine trucks are composed into three platoons by the scheduling optimization algorithm.

The issue lies in optimizing the dispatch of the trucks arriving in period t to form platoons. Specifically, two kinds of costs are included in the objective, i.e., platoon emission reduction Z_1, additional waiting cost Z_2. Among the integer variables, every single variable is deemed to be binary (0-1) variables. Moreover, the following can be specified as:

$$y_{m,k} = \begin{cases} 1, & \text{if the } m\text{-th truck in the } k\text{-th platoon,} \\ 0, & \text{otherwise,} \end{cases} \tag{4}$$

$$y_{m,n,k} = \begin{cases} 1, & \text{if } m\text{-th and } n\text{-th trucks in } k\text{-th platoon,} \\ 0, & \text{otherwise,} \end{cases} \tag{5}$$

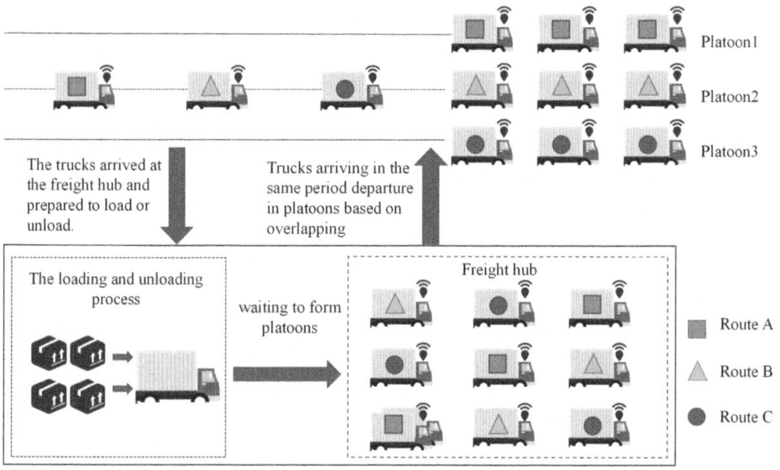

Fig. 2. Operation mode of freight hubs

$\forall m \in [1, N_h - 1], n \in [m+1, N_h], h \in [1, 24]$.

1) Platoon fuel consumption saving Z_1: the energy consumption savings of platoon trucks during driving should be considered, which can be given by

$$Z_1 = \sum_{h=1}^{24} \sum_{k=1}^{K_h} \sum_{m=1}^{N_h-1} \sum_{n=m+1}^{N_h} \sum_{j=1}^{J} [c_d \cdot \frac{1.2 \cdot (\beta_0 - \beta_j)}{10^3} \cdot max(s_{m,n}) \cdot y_{m,n,k}]. \quad (6)$$

2) Waiting cost Z_2: the waiting cost occurs as the trucks are waiting to form a platoon. This includes the driver's salary, the rental cost, and the maintenance cost for the trucks. Therefore, the waiting cost can be computed as

$$Z_2 = \sum_{h=1}^{24} \sum_{k=1}^{K_h} \sum_{m=1}^{N_h-1} c_h \cdot (\tau_h - t_m) \cdot y_{m,k}. \quad (7)$$

According to the previously stated definitions of Z_1, Z_2, Z_1 positive returns, while Z_2 negative costs or losses. Therefore, the objective function can be expressed as $Z = Z_1 - Z_2$, written as

$$\begin{aligned} &\min_{y_{m,k}, y_{m,n,k}} \quad Z \\ &s.t. \quad C_1 : \sum_{m \in N_h} y_{m,k} \leq J, \\ &C_2 : \sum_{k \epsilon K_h} y_{m,k} \leq 1, \quad \forall m \in N_h, \\ &C_3 : y_{m,n,k} \leq \frac{1}{2}(y_{m,k} + y_{n,k}), \quad \forall m \in N_h | n > m, k \in K_h, \\ &C_4 : N_h \leq N_{max}, \quad \forall h \in [1, 24]. \end{aligned} \quad (8)$$

where C_1 stands for the platoon size in h-th time period is restricted, C_2 indicates that the capacity of the freight hub within one time period relies on the ability to receive, process, and dispatch trucks, C_3 implies the m-th truck precedes n in a platoon, C_4 represents that a truck is assigned to no more than one platoon, J is the maximum quantity of trucks in the platoon, N_{max} is the maximum number of trucks permitted in the freight hub.

In order to handle the problem formulated above, we put forward an enhanced mining truck platoons particle swarm optimization (PSO) algorithm with the consideration of ecological benefits and additional costs (including wait costs) to reduce fuel consumption in freight hubs. The details of the designed algorithm as follows: i) It initializes population size, positions, velocities, and parameters. ii) In the main loop, it iterates until reaching the maximum number of generations. iii) In particle iteration, it updates personal best, calculates fitness value, updates velocity and position, adjusts inertia weight, and checks position constraints.

4 Experiment and Result Analysis

4.1 Scenario Design

In this section, the Nguyen and Dupius-based transportation network is adopted for the case study. The main purpose of this case study is to demonstrate the effectiveness of the proposed algorithm. As shown in Fig. 3, this transportation system consists of twelve nodes, seventeen sides, and two OD (origin-destination) pairs. The figures presented on each road's boxes respectively indicate the road length(km) and the free-flow travel time (*hour*). To align with the optimization model, the travel time and distance of each road are reconfigured. Particularly, the length of each path from the origin to the destination is set to be longer than 100 km, and the travel time between the two locations is approximately $3.2h$ to $4.4h$ in this case. Node 1 is utilized as the origin point, also known as the freight hub; Node 2 and Node 3 are chosen as the two destinations; the remaining nodes represent the intersections along the paths. The optimization model schedules the trucks arriving at Node 1 and reaching the destinations via different routes. Suppose 750 trucks carrying three types of goods arrive at the freight hub within a day (24 h), and both the arrival time and destinations follow a random distribution. Table 1 presents the truck flow among Node 1, Node 2, and Node 3 as an example.

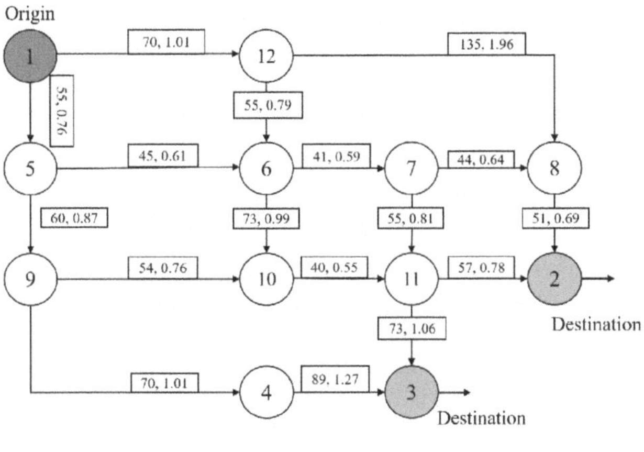

Fig. 3. Nguyen-Dupuis (N-D)-based network

The probability function for a truck's route selection can be determined relying on the classical logit discrete choice theory [18], and it is presented as

$$P^i_{OD,l} = \frac{exp(-\sigma C^i_{OD,l})}{\sum_{OD,l} exp(-\sigma C^i_{OD,l})}, \tag{9}$$

where $P^i_{OD,l}$ represents the probability of choosing route l when delivering good i; σ denotes the scale coefficient; $C^i_{OD,l}$ is the corresponding cost when a truck transports good i via route l.

$$C^i_{OD,l} = \sum_j c^j_e \cdot \alpha^j \cdot L_{OD,l} + c_d \cdot \beta \cdot L_{OD,l} + (c_s + c^i_p \cdot Q) \cdot T_{OD,l}, \tag{10}$$

where c^j_e represents the emission factor (kg/km) of emission j; α^j is the unit social cost ($\$/kg$) for i-th type emission; $L_{OD,l}$ is the driving distance of route l; c_d is the combined cost for fuel consumption ($\$/kg$); β is the fuel consumption factor (kg/km); c_s is the cost for a truck (including the salary of drivers and the rental fee of trucks) ($\$/hour$); c^i_p is the unit delayed penalty cost of good i ($\$/(hour\cdot$ kg)); Q is the capacity of a truck (kg); $T_{OD,l}$ is the free flow driving time of route i. Denote Q_{OD} as the number of trucks from origin O to destination D, thus, the number of trucks on route l can be presented by

$$Q_l = \sum_{OD,\ i} Q_{OD} \cdot P^i_{OD,l} \cdot \tag{11}$$

Table 1. OD design of trucks

NO.	Origin	Destination	Type of good	Trucks	Delay cost
1	Node 1	Node 2	A	150	4
2	Node 1	Node 2	B	200	3
3	Node 1	Node 3	B	150	3
4	Node 1	Node 3	C	250	2

Table 2. Parameters of VSP formulation

Parameter	Value	Parameter	Value
B	0.9872	f_{scale}	17.1
g	$9.8\ m/s^2$	C	0.003616
m	$14\ ton$	C_{D0}	0.75
ρ	$1.206\ \mathrm{kg/m^3}$	C_{D1}	8.125
A	$7.5\ m^2$	C_{D2}	24.744

4.2 Experimental Data and Parameters

The comprehensive dataset of heavy trucks, which contains a total of 75322 s-by-second driving conditions on June 20, 2018, has been adopted for the platoon VSP calculation. This dataset is utilized to represent the specific driving conditions of the lead truck within the platoon. Subsequently, based on the CACC model, the speed and acceleration of the following trucks can be precisely calculated. Correspondingly, the VSP distribution of each platoon truck can also be accurately determined.

For the VSP formulation presented in Eqs. (1) and (2), the parameters associated with heavy trucks are assigned values as indicated in Table 2. The relevant parameters and unit information are clearly defined in the China Traffic Development Annual Report and the China Statistical Yearbook, and it is obtained that Diesel $= 1.1144\ \$/L$. Moreover, the technical data is sourced from official government publications, published papers, and comprehensive surveys [19–21].

4.3 Distribution of Trucks and Goods

The transportation network built upon the framework of Nguyen and Dupius will be brought into play to evaluate the efficiency and functionality of the proffered algorithm. As has been previously expounded, a whopping 750 heavy trucks make their way through the freight hub (Node 1 as illustrated in Fig. 3) on a daily routine. The destinations for these heavy-duty vehicles are randomly selected to be either Node 2 or Node 3. It is strikingly evident that a sum total of 13 routes are available, stretching from the freight hub to the designated destinations. The length of each route and the free-flow travel time, which are related to the road impedance, are different. These 13 routes possess the capability and potential to transport the 750 heavy-duty trucks from the freight hub to the two destinations without any hitch. By relying on Eq. (11), the precise number of trucks on each distinct route, the diverse types of goods being carried, and the essential and fundamental characteristics of each route have been systematically and concisely summarized and presented in Table 3. Since the overlapping sections of the roads and the corresponding lengths of each route have been precisely and accurately established and determined, the overlapping mileage between each route can be meticulously and precisely calculated and evaluated. The detailed and explicit overlapping mileage of these routes is clearly and vividly presented and showcased in Table 4. All in all, there are an aggregate of 78 pairs.

Table 3. Distribution of trucks and goods

OD	No.	Route	Free-flow travel time (h)	Distance	Trucks	Good type
From Node 1 to Node 2	A1	1-12-8-2	3.67	256	50	A
	A2	1-5-6-7-8-2	3.30	236	51	A
	A3	1-5-9-10-11-2	3.72	266	51	B
	A4	1-5-6-7-11-2	3.55	253	49	A
	A5	1-5-6-10-11-2	3.70	270	45	B
	A6	1-12-6-7-8-2	3.73	261	35	B
	A7	1-12-6-10-11-2	4.13	295	38	B
	A8	1-12-6-7-11-2	3.98	278	49	B
From Node 1 to Node 3	B1	1-5-9-13-3	3.91	274	77	C
	B2	1-5-9-10-11-3	4.00	282	71	C
	B3	1-5-6-7-11-3	3.83	269	83	B(43) C(40)
	B4	1-12-6-10-11-3	4.41	311	51	C
	B5	1-5-6-10-11-3	3.98	241	107	B

4.4 Platoon Ecological Characteristics

In this particular subsection, we are going to delve into and explore the interrelationships that exist between the ecological characteristics of the platoon, the distinct departure intervals, and the maximum sizes of the platoon.

Platoon Overlapping Mileage. As illustrated in Fig. 4, We will conduct a detailed discussion on the total overlapping mileage of the 750 trucks that arrive at the freight hub in the context of varying maximum platoon sizes and departure intervals. When the departure interval is set at 15 min, the total overlapping mileage falls within the range of approximately 5489 km to 14654 km. In contrast, when the departure interval is extended to 45 min, the degree of overlap is approximately two times greater than that observed with 15-minute intervals.

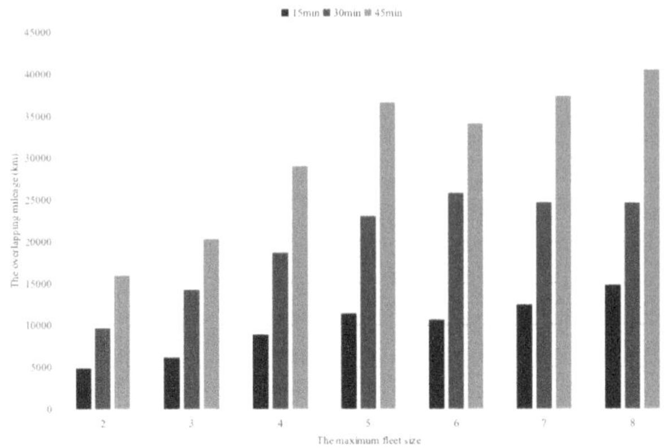

Fig. 4. Overlaps vs the maximum platoon sizes and intervals

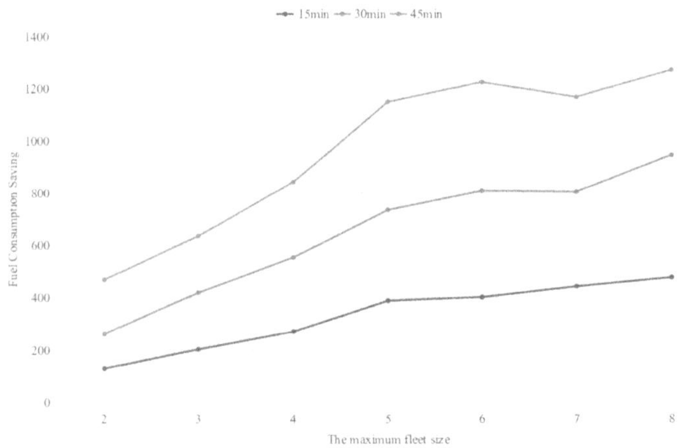

Fig. 5. Fuel consumption saving under various conditions

Aggregate Fuel Consumption. From Table 5, the fuel consumption factors of platoon trucks are computed using Eq. (3). It is clearly observable that the fuel consumption factor of the following trucks is lower than that of the trucks in the front. For instance, when driving within the speed range of [60–80)km/h, the consumption factor of the 8th truck is 14.5% less than that of the 1st truck.

The influence of maximum platoon size and departure interval on fuel consumption is clearly demonstrated in Fig. 5. It can be observed that the CACC-platoon trucks are capable of saving fuel consumption by approximately 128 L–1260 L. The trend of this reduction is similar to that witnessed in emission reduction. Specifically, when the departure interval increases, the amount of fuel consumed is significantly lower. In conclusion, the proposed algorithm is capable of reducing a considerable quantity of fuel consumption for long-distance transportation.

Table 4. The overlapping mileage of routes (km)

Route No.	Overlap	Route No.	Overlap	Route No.	Overlap	Route No.	Overlap
A1 & A2	51	A4 & A7	57	A1 & B3	0	A5 & B3	100
A1 & A3	0	A4 & A8	153	A1 & B4	70	A5 & B4	113
A1 & A4	0	A5 & A6	0	A1 & B5	0	A5 & B5	213
A1 & A5	0	A5 & A7	170	A2 & B1	55	A6 & B1	0
A1 & A6	121	A5 & A8	57	A2 & B2	55	A6 & B2	0
A1 & A7	70	A6 & A7	125	A2 & B3	141	A6 & B3	41
A1 & A8	70	A6 & A8	166	A2 & B4	0	A6 & B4	125
A2 & A3	55	A7 & A8	182	A2 & B5	100	A6 & B5	0
A2 & A4	141	B1 & B2	115	A3 & B1	115	A7 & B1	0
A2 & A5	100	B1 & B3	55	A3 & B2	209	A7 & B2	40
A2 & A6	136	B1 & B4	0	A3 & B3	55	A7 & B3	0
A2 & A7	0	B1 & B5	55	A3 & B4	40	A7 & B4	238
A2 & A8	41	B2 & B3	128	A3 & B5	95	A7 & B5	113
A3 & A4	112	B2 & B4	113	A4 & B1	55	A8 & B1	0
A3 & A5	152	B2 & B5	168	A4 & B2	55	A8 & B2	0
A3 & A6	0	B3 & B4	73	A4 & B3	196	A8 & B3	96
A3 & A7	97	B3 & B5	128	A4 & B4	0	A8 & B4	125
A3 & A8	57	B4 & B5	186	A4 & B5	100	A8 & B5	0
A4 & A5	157	A1 & B1	0	A5 & B1	55		
A4 & A6	41	A1 & B2	0	A5 & B2	55		

Table 5. Fuel consumption factors of different trucks in platoon (g/km)

Speed range (km/h)	1st truck	2nd truck	3rd truck	4th truck	5th truck	6th truck	7th truck	8th truck
[0–20)	521.885	481.699	466.933	457.340	450.508	446.323	444.058	442.249
[20–40)	400.016	369.615	359.155	348.739	344.262	341.594	339.629	337.965
[40–60)	240.525	223.207	214.926	209.575	206.678	204.738	203.565	202.618
[60–80)	193.566	180.036	174.617	170.915	168.775	167.253	166.311	165.583

As presented in Fig. 6, we delve into the overall economic benefits of the platoon under varying departure intervals and platoon sizes, as well as the optimal scheme. In this article, it is determined that the optimal departure break is 30 min and the allowable platoon size is 5. Based on the proposed algorithm, the total profitability reaches its peak at 512.7$.

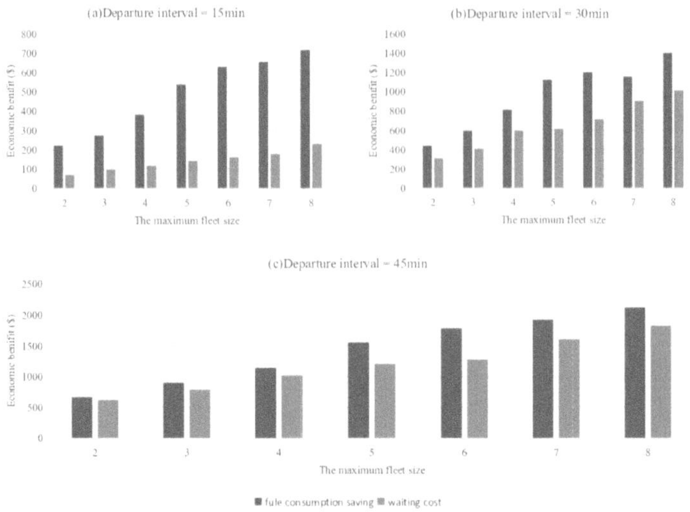

Fig. 6. Benefits and costs of different aspects

Deploying the proposed truck platooning optimization algorithm in real-world freight hubs would incur several additional costs that should be considered for a comprehensive evaluation of its feasibility and practical impact: **1)Infrastructure Costs:** V2V Communication Equipment: Each truck would need to be equipped with V2V communication devices compatible with 6G networks. This would involve significant upfront costs for hardware installation and ongoing maintenance. Upgrading or installing 6G network infrastructure along transportation routes to ensure consistent connectivity for the platoons. **2)Software Development and Integration:** Customization of the algorithm for specific freight hub operations and integration with existing logistics management systems and development of user interfaces for dispatchers and drivers. **3)Operational Adjustments:** Potential redesign of freight hub layouts to accommodate platoon formation and dissolution; adjustments to scheduling and routing processes to align with the new platooning system. While these additional costs are significant, they should be weighed against the long-term benefits of reduced fuel consumption, lower emissions, and improved operational efficiency. A thorough cost-benefit analysis, considering both immediate implementation costs and long-term savings, would be crucial for stakeholders to make informed decisions about adopting this technology. Additionally, as the technology matures and becomes more widespread, some of these costs may decrease, potentially improving the overall economic viability of the system.

5 Conclusion

In this paper, we propose an enhanced particle swarm optimization algorithm for medium-sized truck convoys at freight hubs, considering ecological benefits and formation costs, and utilizing a CACC model to control the fleet. This model is validated on an improved Nguyen and Dupis transport network. The study reveals: 1) VSP values based on actual heavy truck data aid in calculating fuel consumption; 2) as the maximum fleet size increases, the trend of reduction in fuel consumption slows; 3) Fuel consumpution saving and additional waiting costs rise with increasing departure intervals and maximum fleet size. Future research directions include: 1) fleets can depart but not join midway; 2) exploring fixed departure intervals, dynamic scheduling might be more realistic.

Acknowledgment. This work is supported in part by Science and Technology Innovation Key R&D Program of Chongqing (Grant No. CSTB2023TIAD-STX0035), National NSFC (Grants No. 62372072, 62102053, 62072060, 62067206 and 61972222), Chongqing Research Program of Basic Research and Frontier Technology (Grant No. cstc2022ycjhbgzxm0058), Haihe Lab of ITAI (Grant No. 22HHXCJC00002), the Natural Science Foundation of Chongqing, China (Grant No. CSTB2022NSCQ-MSX1104), Key Laboratory of Big Data Intelligent Computing, Chongqing University of Posts and Telecommunications (Grant No. BDIC-2023-B-003), Regional Innovation Cooperation Project of Sichuan Province (Grants No. 2023YFQ0028, 24QYCX0019), and Regional Science and Technology Innovation Cooperation Project of Chengdu City (Grant No. 2023-YF11-00023-HZ), and Fundamental Research Funds for the Central Universities (Grant No. 2024CDJGF-003).

References

1. Barua, L., Zou, B., Choobchian, P.: Maximizing truck platooning participation with preferences. Transp. Res., Part E. Logistics Transp. Rev. **179**, 1366–5545 (2023)
2. Feng, R., et al.: A comparative study on the energy flow of a hybrid heavy truck between AMT and MT shift mode under local driving test cycle. Energy Convers. Manage. **256**, 115359 (2022)
3. Huang, Z., et al.: A mixed-bouncing based non-stationarity and consistency 6G V2V channel model with continuously arbitrary trajectory. IEEE Trans. Wireless Commun. (2023)
4. Li, W., Meese, C., Guo, H., Nejad, M.: Aggregated zero-knowledge proof and blockchain-empowered authentication for autonomous truck platooning. IEEE Trans. Intell. Transp. Syst. (2023). https://doi.org/10.1109/TITS.2023.3271436
5. Chen, X., Leng, S., He, J., Zhou, L., Liu, H.: The upper bounds of cellular vehicle-to-vehicle communication latency for platoon-based autonomous driving. IEEE Trans. Intell. Transp. Syst. (2023). https://doi.org/10.1109/TITS.2023.3263239
6. Guo, C., Fu, C., Luo, R., Yang, G.: Energy-oriented car-following control for a front-and rear-independent-drive electric vehicle platoon. Energy **257**, 124732 (2022)
7. Mahdinia, I., Arvin, R., Khattak, A.J., Ghiasi, A.: Safety, energy, and emissions impacts of adaptive cruise control and cooperative adaptive cruise control. Transp. Res. Rec. **2674**(6), 253–267 (2020)
8. Wang, C., Gong, S., Zhou, A., Li, T., Peeta, S.: Cooperative adaptive cruise control for connected autonomous vehicles by factoring communication-related constraints. Transp. Res. Part C: Emerg. Technol. **113**, 124–145 (2020)

9. Chen, S., Wang, H., Meng, Q.: Autonomous truck scheduling for container transshipment between two seaport terminals considering platooning and speed optimization. Transp. Res. Part B: Methodol. **154**, 289–315 (2021)
10. Xu, M., Yan, X., Yin, Y.: Truck routing and platooning optimization considering drivers' mandatory breaks. Transp. Res. Part C: Emerg. Technol. **143**, 103809 (2022)
11. Xie, S., Lang, K., Qi, S.: Aerodynamic-aware coordinated control of following speed and power distribution for hybrid electric trucks. Energy **209**, 118496 (2020)
12. Milanés, V., Shladover, S.E.: Modeling cooperative and autonomous adaptive cruise control dynamic responses using experimental data. Transp. Res. Part C: Emerg. Technol. **48**, 285–300 (2014)
13. Xiao, L., Wang, M., Schakel, W., van Arem, B.: Unravelling effects of cooperative adaptive cruise control deactivation on traffic flow characteristics at merging bottlenecks. Transp. Res. Part C: Emerg. Technol. **96**, 380–397 (2018)
14. Jimenez-Palacios, J.L.: Understanding and quantifying motor vehicle emissions with vehicle specific power and TILDAS remote sensing. Massachusetts Institute of Technology (1998)
15. Song, M., Chen, F., Ma, X.: Organization of autonomous truck platoon considering energy saving and pavement fatigue. Transp. Res. Part D: Transp. Environ. **90**, 102667 (2021)
16. Hucho, W., Sovran, G.: Aerodynamics of road vehicles. Annu. Rev. Fluid Mech. **25**(1), 485–537 (1993)
17. Jimenez, J.L., McClintock, P., McRae, G., Nelson, D.D., Zahniser, M.S.: Vehicle specific power: a useful parameter for remote sensing and emission studies. In: Ninth CRC On-Road Vehicle Emissions Workshop. San Diego, CA (1999)
18. Cooper, A.B., Millspaugh, J.J.: The application of discrete choice models to wildlife resource selection studies. Ecology **80**(2), 566–575 (1999)
19. NBSC, "National data." https://data.stats.gov.cn/easyquery.htm?cn=A01
20. BMCDR. "Beijing electric vehicles charging infrastructure special plan of 2016-2020." (2016) http://www.evpartner.com/news/12/detail-19416.html
21. Song, S.: Transport emissions & social cost assessment: Methodology guide" (2017)

Artificial Intelligence and Digital Economy

A Software-Hardware Co-design Platform for Convolutional Neural Network Accelerator with Tengine

Zhiwei Liu(✉), Shangwen Sui, Wanyu Lu, Bin Yu, and Ao Li

School of Computer Science and Technology, Harbin University of Science and Technology, No. 52 Xuefu Road, Harbin 150080, Heilongjiang, China
zwliu@hrbust.edu.cn

Abstract. With the development of convolutional neural network, its model scale and data volume are getting bigger and bigger, so the convolutional neural network gas pedal has been born. NVDLA as an open-source project from NVIDIA, supports a variety of classical models. However, due to the complexity of the ASIC design flow and the lack of support for the latest operators, its application scenarios are significantly limited. In this paper, we propose a method to map NVDLA onto FPGA, and employ the Tengine toolchain to replace the original software stack provided by the NVDLA. This approach optimizes the performance of the supported LeNet5 and ResNet18 models. Experimental results show that the Tengine toolchain increases inference speed by 151% compared to the NVDLA website's software toolchain.

Keywords: NVDLA · Tengine · MPSoC

1 Introduction

As the development of deep learning in the field of target detection becomes more and more mature [1], its accuracy and robustness are getting better and better. However, along with the maturity of the technology, the size of its model is also getting bigger and bigger. When we deploy deep learning model inference, it is mainly deployed on three kinds of hardware devices: central processing unit, graphical processing unit and specialized inference gas pedal unit [2]. The central processing unit, as the main computing unit of the computer, has the best versatility, but its excessive instruction control leads to the fact that it is not fully capable of such data-intensive tasks as deep learning reasoning; the graphical processing unit has excellent computational performance, which has a wide range of applications in deep learning, but the excellent computational performance is built on the premise of not taking into account the cost of computation, the cost of arithmetic power, and the heat dissipation of the equipment. Therefore, it is often deployed in the cloud for use as a training server, and often not deployed in embedded devices for inference use. Dedicated inference gas pedals have the advantage of high performance and low power consumption, and thus are most suitable for deployment on embedded platforms [3].

X. Li et al. (Eds.): 6GN 2024, LNICST 615, pp. 169–178, 2025.
https://doi.org/10.1007/978-3-031-93193-2_14

Due to the complexity of deep learning models, dedicated inference gas pedals in the deployment of embedded platforms, we usually do not use the hardware gas pedal alone, but design it to be used as a heterogeneous SoC system, which contains not only the design of the hardware, but also the design of the software model compiler, the design of the workflow scheduling, and the amount of work and workload requires a mature team with a certain size to Completed, but after analyzing the current deep learning model improvements, it was found that the current deep learning model improvements in the field of target detection are often in the structure, in essence, its computational model has not changed, for example, convolutional neural network, the basic structure of its convolution, activation, pooling, and full connectivity has not changed, and for an inference gas pedal that is already mature, want to apply the new model For a mature inference accelerator, the most important work is the adaptation of the software model compiler and the design of workflow scheduling for the new model.

For the problem of adapting new operators for new models, the Thinker inference gas pedal designed in Research [4] implements a platform computation using the Thinker gas pedal based on Coarse Grain Reconfigurable Array (CGRA) arrays; the hardware acceleration methods such as parallel combination strategy, data fixed-point quantization, and multi-channel data transfer used in Research [5] design a hardware acceleration method for the YOLOv4-tiny model of target detection gas pedal architecture; Research [6] used the method based on row-fixed data flow structure and adding control circuits in PE to skip unnecessary computations to adapt the new model, Research [7] designed an FPGA-based GEMM general-purpose matrix processor to accelerate the YOLOv3-tiny model and reduced the accuracy of the original model to reduce the memory space and bandwidth requirements; Research [8] proposed a novel gas pedal architecture, AI-MT, to support a wide variety of models by dividing each layer into multiple identical sub-layers creating fine-grained tasks; Research [9] proposed a scalable inference acceleration processor DSIP where the main processor is responsible for sending and receiving feature maps in order to process the neural network in a scalable way. But these are not open source and it is difficult to study them in depth, but NVDLA [10] is an open source neural network gas pedal with software environment support.

NVDLA is a deep learning gas pedal released by NVIDIA in 2017 as an open source, configurable, fixed-function deep learning gas pedal. It functions to perform typical convolutional class deep learning computations. The modular architecture is scalable, reconfigurable, and has good portability. However, due to the extremely slow update and maintenance speed of the NVDLA community, many mainstream target detection inference frameworks are no longer supported, and NVDLA currently only supports the compilation of the Caffe framework, resulting in it being no longer practical as an engineering project. Based on this, this paper proposes a method to extend the functionality of NVDLA, adopting the heterogeneous SoC design method of NVDLA + ARM, solving the problem of only supporting Caffe models by using the open-source reasoning tool Tengine, and solving the problem of the limited support of operators by NVDLA, and finally, developing a method based on the above to realize the target detection of tomato using the ZCU104 development board application.

2 Introduction to the Hardware and Software Platforms

2.1 Introduction to NVDLA

NVDLA is a deep learning model inference gas pedal open-sourced by NVIDIA in 2017 [11], which can be used for inference acceleration of convolutional neural network models (CNN). A total of three scale sizes are currently released, respectively small, large, and full versions, of which the small version takes up the smallest resources and only supports INT8 data types, which is suitable for deployment in embedded devices.

Deploying NVDLA consists of two major parts: hardware design includes packaging the NVDLA acceleration core into IP cores and designing the SoC system containing the NVDLA IP cores, and software design includes the compiler compiler that compiles and parses the CNN models, the process of converting the models of various front-end software into a file that can be understood by the hardware, and the runtime part that allocates scheduling time for the accelerated models. The software design includes a compiler that compiles and parses the CNN models, a runtime component that allocates and schedules the models to be accelerated. The overall structure of the NVDLA project is analyzed in order to adapt new models to it.

2.2 Introduction to the NVDLA Hardware Section

Its hardware structure mainly contains six main modules, including all the logic circuit structure of the NVDLA inference acceleration part, the convolutional kernel module is the core of the acceleration part, which is composed of MAC arrays and mainly accelerates the convolutional computation; the single data processor module mainly supports the activation function; the planar processor module mainly supports pooling operations; the channel data processor module supports the normalization module; and there are dedicated storage and data reorganization engine and bridge DMA module. The channel data processor module accelerates the normalization module, as well as the dedicated storage and data reorganization engine and the bridge DMA module. The dedicated storage and data reorganization engine and the bridge DMA module are not supported in the small version. The NVDLA hardware architecture is shown in Fig. 1.

The structure of the small version of the NVDLA-IP core is shown in the figure, including the CSB interface, IRQ interface and DBBIF interface. The CSB interface is connected to the central processor as a configuration bus interface to configure the NVDLA internal registers and assign workflows; the DBBIF interface is connected to the DRAM as a data backbone interface to transfer access data from NVDLA; and the IRQ interface is an external interrupt interface to transfer interrupt signals. In the SoC design, the central processor is also connected to the DRAM to obtain operation instructions.

2.3 Introduction to the NVDLA Software Section

The software part consists of two parts, the compiler and runtime, As shown in Fig. 2 [12], which are both running in the central processor of the SoC system. The compiler, i.e., the convolutional neural network compiler part, the main process is to firstly transform the trained caffe model into the structure of the computational graph, then optimize the

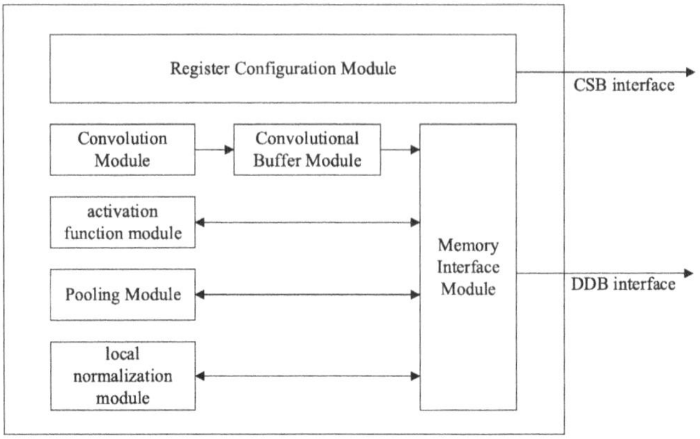

Fig. 1. NVDLA Hardware Architecture Diagram

network structure of the computational graph, and then transform it into the structure of the abstract syntax tree adapted to the NVDLA. The runtime consists of two parts, UMD and KMD, the UMD is the upper application driver, which is responsible for receiving and parsing the loadable file and finally sending it to KMD; the KMD is the kernel driver, which is the driver with the closest relationship with the hardware, and is responsible for taking the inference model received and parsing it to KMD. It is responsible for sending the received hardware configuration information in the inference model to the configuration registers in the hardware structure, reading the data from the memory, executing the allocation of the registers and the execution of the tasks, and finally outputting the results. However, because NVDLA only supports models trained by caffe framework, many other mainstream frameworks are not supported, such as pytorch, tensorflow, etc., so it greatly limits the use of NVDLA scenarios, to address this problem, we introduce another open source tool - Tengine.

2.4 Introduction to Tengine

Tengine is a lightweight edge AI inference tool with a high degree of flexibility [13], which can quickly deploy the trained model structure into various devices, including CPU, GPU and NPU, and also has a compiler that includes support for a variety of models, as well as model transformation, model quantization, and a variety of side-end deployments, which can efficiently leverage the heterogeneous arithmetic on a variety of devices. Supports most modeling frameworks, including pytorch, tensorflow, caffe, onnx, and so on. If there is a need for quantization, you can use Tengine's own quantization tool to perform symmetric and asymmetric quantization of int8, and directly import the network model and training set to complete the quantization. When deployed at the backend, Tengine has a unique automatic graph-cutting mechanism, which can run the operators that support npu on npu, and cut the operators that don't support npu to cpu or other devices, which gives full play to the heterogeneous computing capability of the devices. This design uses Tengine compiler instead of the official compiler tool to adapt to the mainstream model.

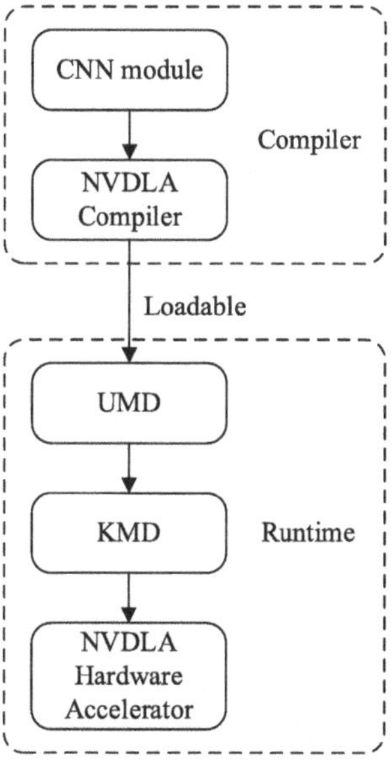

Fig. 2. NVDLA Software Flowchart

3 Methodology

3.1 Hardware Solutions

SoC Architecture Design. The SoC architecture design in this paper focuses on the experimental development platform, bus design and clock design.

Experimental Development Platform. The experimental platform selected for this design is the ZCU104, a development board based on the Xilinx Zynq UltraScale and MPSoC, which is equipped with an FPGA chip of the model XCZU7EV, which contains hardware resources as shown in the figure, and two ARM core processors, one is a quad-core Arm Cortex-A53 processor, suitable for running complex operating systems and applications, and a dual-core Arm Cortex-R5 processor for real-time tasks and control applications. The SoC system can be developed and designed on a PC using Vivado and Petalinux software.

Bus Design. In the design of this paper, a Linux kernel runs on the main processor to manage the hardware gas pedal and related tasks. The controller below the CPU is responsible for scheduling and fine-grained configuration of the hardware gas pedal, but the host will still be responsible for coarse-grained scheduling on the hardware gas pedal, memory access to the hardware gas pedal, IOMMU mapping, memory allocation of the hardware gas pedal's input data and the fixed-weight array, and synchronization between other system components and tasks running on the gas pedal. Synchronization between other system components and tasks running on the accelerator. The various peripherals and NPU configuration ports are interconnected via an APB standard protocol low-speed bus interface with 32-bit data bit width and operating frequency up to 100 MHz. In the system console, the necessary peripheral ports, such as SDIO, UART, GEM, etc., also need to be configured.

Clock Design. This soc clock source includes a 32.768 kHz passive crystal oscillator, a 33.33 MHz active crystal oscillator and a 200 MHz differential clock source. The passive crystal oscillator mainly provides clock signals for Timer, while the active crystal oscillator provides clock signals for the low-speed devices and the CSB interface of NPU through PLL phase-locked loops after multiplication, division and modulation, and the CSB interface of NPU is provided directly by the active crystal. The CSB interface of NPU provides the clock signal for the active crystal directly, after the active crystal is multiplied 90 times, it is divided into SD Controller after 8 divisions, SPI after 75 divisions, UART after 15 divisions, PCIe after 6 divisions, and USB after 6 divisions. The differential clock source mainly provides the clock signal for the DDR and the DDB interface of NPU, and the DDB interface of NPU provides the clock signal for the DDR and the DDB interface of NPU. The DDB interface of NPU provides clock signals for DDR and DDB interface of NPU, and the differential clock source provides clock signals for DDR after 6 times frequency of the differential signals directly.

Determination of NVDLA Configuration Parameters. The small version of NVDLA has 64 MAC array units that support INT8, while the large and full versions have 2048 MAC array units that support multi-precision. For most development boards, the on-board resources can only support the small version of NVDLA.

3.2 Software Programs

First of all, this paper is designed to port Ubuntu16.04 system on ARM core, and complete all user layer operations on this Linux kernel system. The software design includes Compiler and Runtime, the software supporting the NVDLA official website is currently only adapted to Lenet, GoogleNet, ResNet, etc., and is not adapted to some other mainstream models at present, but the software layer of NVDLA, no matter it is the compiler or the runtime, provides a portable layer, this paper will be the NVDLA in the In this paper, the software layer in NVDLA is ported to Tengine by packaging it into an API interface, and the reasoning operation is accomplished using the combination of NVDLA and Tengine.

3.3 Design Process

Hardware Platform. The NVDLA community has released related projects for ASIC designs. If you want to map NVDLA on FPGA, the first thing you need to consider is its RAM for ASIC design. The RAM designed on the official website is an RTL behavioral level netlist file, so if you map it directly on FPGA, FPGA will waste the on-chip BRAM resources by implementing its function through LUT. The second thing to consider is the convolutional kernel module. The convolutional array module on NVDLA is realized by a two-dimensional MAC array, where each MAC unit performs a multiply-add operation, and this computation mode just makes use of the FPGA's resources. The last thing to consider is that there is a clock tree design in the ASIC design, but another clock tree will be automatically generated when the FPGA project is packaged, and the two clock trees will lead to timing conflicts and time violations.

When packaging the NVDLA IP core, based on the comprehensive consideration of the interfaces provided by the official website and the interfaces required by the SoC, five control bus interfaces are encapsulated for the control bus protocol: operating clock, control bus clock, event interrupt, control bus, and external memory bus. A picture of the packaged IP core is shown in Fig. 3. Among them, 4KB of addressing space needs to be reserved in the control bus.

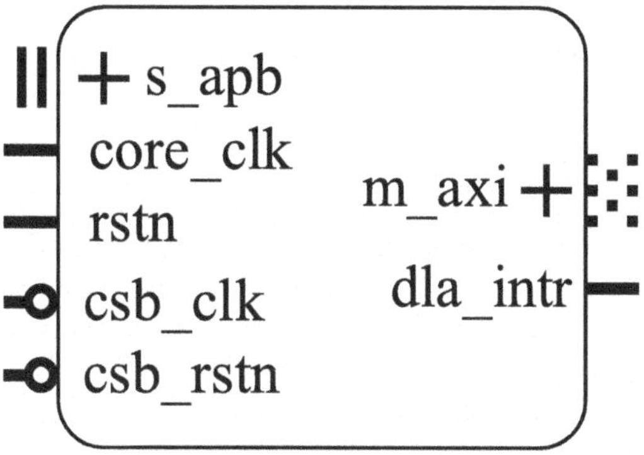

Fig. 3. NVDLA-IP core picture

When designing the SoC in ZYNQ platform, it is divided into PL side and PS side, PL side is the programmable logic control side, which is the FPGA platform part; PS side is the system processing side. In this SoC design, the NVDLA IP core uses the AXI protocol to transfer the data backbone to the PS side, and the APB protocol to transfer the control information. Firstly, the hardware is initialized, and then it starts to transfer the data; the central processor transfers the data to the DDR, and the gas pedal reads the data from the DDR directly through the DMA; at the end of the transfer the DMA After the transfer, the DMA sends an interrupt to the central processor; the central processor sends configuration information to the NVDLA, including register configuration and the workflow of each module; after the end of the task or the occurrence of a bug, it sends an end interrupt to the central processor.

In the PS side, you also need to configure the necessary peripheral ports such as SDIO, UART and GEM. The SoC design architecture is shown in Fig. 4.

Software Platforms. After the hardware platform is built again, the next need to port the PS side of an operating system platform that provides software execution and KMD construction, ZYNQ provides Petalinux development kit, an operating system for the linux kernel, but because of its development kit provides less, so this design is only used as a KMD program generation, PS side of the operating system used for the ubuntu16.04. In the process of using Petalinux to generate KMD, we only need to add the clock tree of the corresponding board and the NVDLA/HW code from the official website in the project to generate KMD.

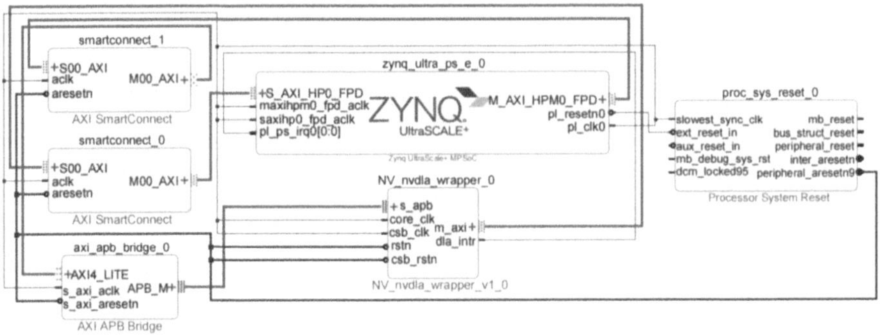

Fig. 4. Hardware SoC Design Architecture Diagram

After entering the operating system, download the NVDLA/SW code from the official website and the Tengine source files, compile them to generate the libcompiler library and libruntime library, and put them into the Tengine project, so that the original NVDLA UMD can be called as an API in Tengine, and pay attention to the inclusion of the In the compilation process of the project, be careful to include the NVDLA part, otherwise, the reasoning will be done by default using the central processor. The NVDLA acceleration platform is now complete.

NVDLA only supports INT8 data models, but the mainstream models are trained on GPUs. Since GPUs are basically designed with large bit widths, the mainstream models are in Float32 format, and most of the models are not structurally adapted to NPUs, so structural optimization and quantization of the models have become a necessary part of the process, and Tengine provides a corresponding tool. For convolutional neural network models, there are a large number of structural blocks of convolutional and Batch-Normlization operators, both of which belong to linear operations, and fusion can significantly reduce the number of network layers, but also reduce the acceleration cache and data access to the computational unit.

For quantization, Tengine adopts a linear quantization scheme, in which the image size and quantization strategy are formulated, and the quantization can be completed by extracting about 800 pieces of data from the dataset.

4 Experimental Results

4.1 Selection of Test Platform

In this paper the hardware design is developed in VIVADO software for NVDLA IP core and SoC system and petalinux software is used for KMD development. The test model is trained using PC, xxx model, xxx model and xxx model are trained using coco training set, the model is optimized and quantized using Tengine (X86 version) in linux VM, UMD is compiled on the development board, Tengine (ARM64) is compiled on the development board for inference, and finally the experimental results are analyzed.

4.2 Experimental Comparison

Use the official NVDLA compiler and Tengine compiler to optimize the above two models and quantize them in INT8 format respectively, input the images of the test set and test them on CPU, GPU and hardware gas pedal respectively to get their test time and power consumption, compare them and the results are shown in the Table .1.

Table 1. Model inference results

	CPU	NVDLA	NVDLA + Tengine
LeNet5(ms)	15.74	11.40	12.57
ResNet18(ms)	145.21	40.26	15.98

As can be seen from the graph, due to the relatively simple structure of the LeNet5 network, the optimization effect is not obvious. For the ResNet18 network, due to its complex structure, the optimized model after Tengine's optimization has a significant improvement in reasoning efficiency, with a speed increase of 151% compared to the official NVDLA compiler, and a speed increase by 8 times compared to the CPU.

5 Conclusion

In this paper, we propose an efficient mapping method on FPGA by deeply analyzing NVDLA from the hardware side to the software side, and use the Tengine open-source tool to solve the problem that NVDLA can not be adapted to new convolutional neural network models. Another SoC design on the ZYNQ platform is proposed and implemented as a heterogeneous deep learning model acceleration platform and deployed on the ZCU104 development board. Experiments are conducted to accelerate three CNN models, LeNet5 and ResNet18, and the experimental results show that the gas pedal acceleration effect is obvious compared with the CPU, and the inference speed of the optimized model is obviously larger than that of the unoptimized model, especially the model with complex structure.

Acknowledgement. This work was supported by the National Key Research and Development Program of China (Grant No. 2022YFD2000500).

References

1. Pan, S.B., et al.: Requirements and policies of cloud computing security. In: National Academic Conference on Communication Security, pp. 152–156 (2010)
2. Shen, Y.M., Ferdman, M., Milder, P.: Maximizing CNN accelerator efficiency through resource partitioning. In: Proceedings of the 44th Annual International Symposium on Computer Architecture (ISCA), pp. 535–547. IEEE, Toronto, Canada (2017)

3. Mao, Y.Y.: Design methodologies for green communications and mobile-edge computing systems in 5G networks. Phd thesis. Hong Kong university of science and technology, Hong Kong (2017)
4. Yin, S., et al.: A high energy efficient reconfigurable hybrid neural network processor for deep learning applications. IEEE J. Solid-State Circuits **53**(4), 968–982 (2017)
5. Zhao, V., Zhao, Y., Jin, Y., Wang, Y., Gang, Y., Wang, Y.: YOLOv4-tiny model for accelerated design of edge computing platform. Internet Things Technol. **14**(01), 93–97 (2024)
6. Chen, Y.-H., Emer, J., Sze, V.: Eyeriss: a spatial architecture for energy-efficient dataflow for convolutional neural networks. In: ACM/IEEE 43rd Annual International Symposium on Computer Architecture (ISCA), pp. 367–379 (2016)
7. Shawahna, A., Sait, S.M., El-Maleh, A.: FPGA-based accelerators of deep learning networks for learning and classification: a review. IEEE Access **7**, 7823–7859 (2018)
8. Baek, E., Kwon, D., Kim, J.: A multi-neural network acceleration architecture. In: Proceedings of the 2020 ACM/IEEE 47th Annual International Symposium on Computer Architecture (ISCA), pp. 940–953. IEEE (2020)
9. Jo, J., Cha, S., Rho, D., Park, I.C.: DSIP: a scalable inference accelerator for convolutional neural networks. IEEE J. Solid-State Circuits **53**(2), 605–618 (2017)
10. Mittal, S.: A survey of FPGA-based accelerators for convolutional neural networks. Neural Comput. Appl. **32**(4), 1109–1139. Springer (2020)
11. NVIDIA Corporation. Hardware manual. [EB/OL]. http://nvdla.otg/hw/contents.html. Accessed 02 July 2021
12. Li, M., et al.: The deep learning compiler: a comprehensive survey. IEEE Trans. Parallel Distrib. Syst. **32**(3), 708–727 (2021)
13. Xiong, A.: OPEN AI LAB: Empowering AIoT with platform thinking. Robot. Ind. **2020**(05), 71–76 (2020)

Personalized Federated Learning Algorithm Based on Shuffle Differential Privacy

Ruifeng Zhao[1], Jiangang Lu[1], Zufeng Hou[2], Yong Zhang[2], Chao Wang[2], Jiantian Chen[2], Kaiyan Pan[3(✉)], and Haixin Liu[3]

[1] Electric Power Dispatching and Control Center of Guangdong Power Grid Co., Ltd., Guangzhou, China
lujiangang@gddd.csg.cn

[2] Zhuhai Power Supply Bureau, Guangdong Power Co., Ltd., Zhuhai, China

[3] Dongfang Electronics Cooperation, Yantai, China
kaiyanpan@126.com

Abstract. Federated learning, as an innovative machine learning technique, not only effectively protects user data privacy and enhances data efficiency but also supports collaboration among multiple participants, demonstrating its extensive application potential. Nevertheless, federated learning still faces some limitations in practical applications. Specifically, when client data are non-independent and identically distributed (Non-IID), the model's generalization performance and convergence speed are significantly affected. Additionally, the gradients obtained by users during local model training may be reconstructed by attackers to reveal the complete training data. Addressing these issues of poor model generalization in Non-IID data scenarios and potential privacy leakage during gradient uploads in federated learning, this paper proposes a personalized federated learning algorithm, SS-PFL, based on shuffled differential privacy. The algorithm initially adopts a bi-level optimization approach to train personalized models locally on clients, optimizing local model performance. Secondly, it introduces a shuffled server to ensure secure aggregation of gradient parameters within a unified privacy budget scenario. Finally, experiments are conducted to verify whether SS-PFL can effectively resist reconstruction attacks while improving accuracy.

Keywords: Federated learning · Data privacy · Reconstruction attack · Differential privacy

1 Introduction

Federated learning is an emerging machine learning approach that allows multiple devices to train models locally and upload model parameters to a server

X. Li et al. (Eds.): 6GN 2024, LNICST 615, pp. 179–190, 2025.
https://doi.org/10.1007/978-3-031-93193-2_15

for aggregation [9]. After aggregating the model updates, the server sends the aggregated information back to the participating clients for further training iterations until a global model with sufficient accuracy is obtained. According to the distribution of data among participants, federated learning can be classified into three types: horizontal federated learning, vertical federated learning, and federated transfer learning, which correspond to scenarios where the label spaces overlap, the sample spaces overlap, and both the label spaces and sample spaces have minimal overlap, respectively [12].

Consequently, federated learning still faces several challenges in practical applications:

- The non-independent and identically distributed (non-IID) data. In federated learning scenarios, data is stored only on local clients, which leads to the non-IID data problem. Non-IID data in federated learning can be classified into several categories: feature distribution skew, label distribution skew, concept shift (where data from different clients has the same labels but different features, or the same features but different labels), and data skew or imbalance. Traditional federated averaging algorithms exhibit slow convergence and lower model accuracy when faced with non-IID data.
- Privacy protection. While data does not leave the local environment during the federated learning training process, it introduces a significant amount of parameter exchange (gradient uploads and model downloads). Existing research has demonstrated that malicious training parties can reconstruct complete training data from explicit gradients. Therefore, there is an urgent need for more measures to protect the data privacy of training parties.

Addressing the aforementioned issues, to ensure model performance while protecting the data privacy of training parties, this chapter proposes a personalized federated learning algorithm based on shuffled differential privacy, named SS-PFL (Secure Shuffling Personalized Federated Learning). Firstly, a bi-level optimization method is employed to train personalized models locally on clients, thereby optimizing the performance of local models. Secondly, by introducing a shuffling server, secure aggregation of gradient parameters is ensured under a unified privacy budget scenario. Experimental results demonstrate that, compared to existing privacy-preserving federated learning methods, this algorithm not only achieves improved classification accuracy in multiple scenarios but also effectively defends against reconstruction attacks.

The remainder of this paper is organized as follows. Section 2 introduces related works. Section 3 briefly describes related preliminaries contained in our algorithm. Section 4 presents the system model. Section 5 shows the overall detailed flow of SS-PFL. Section 6 present theperformance analyses. Finally, conclusions are drawn in Sect. 7.

2 Related Works

Federated learning, as a privacy-preserving distributed machine learning paradigm, achieves a certain level of data privacy protection by training models on local devices and sharing only model updates rather than raw data. However, sharing model parameters alone cannot fully guarantee data privacy. The DLG attack algorithm proposed by Zhu *et al.* [9] demonstrates that attackers can easily infer private data from shared model information. To further enhance the privacy protection capabilities of federated learning, researchers have considered the following aspects:

Differential Privacy (DP): Differential privacy is a commonly used privacy protection framework that introduces random noise into shared data, making it difficult for attackers to distinguish between two similar datasets, thereby protecting individual privacy. In federated learning, differential privacy mechanisms can be introduced at various stages, such as adding noise during model training, model aggregation. Representative works include DP-FedAvg [13].

Secure Multi-Party Computation (MPC): Secure multi-party computation [1] allows multiple participants to jointly compute a function without revealing their respective private data. In federated learning, MPC can be used to achieve secure model aggregation, so that the server cannot obtain the local model parameters of each client but only the encrypted aggregated results. Representative works include SecureAgg [2], SecureBoost [4], etc.

Homomorphic Encryption (HE): Homomorphic encryption [8] is an encryption technique that allows direct computation on encrypted data. In federated learning, homomorphic encryption can be used to achieve privacy-preserving model aggregation and prediction. Each client sends encrypted local model parameters to the server, and the server performs aggregation in an encrypted state to obtain the encrypted global model. Representative works include FedHE [3].

Local Differential Privacy (LDP): Different from traditional differential privacy, local differential privacy [11] perturbs data at the data source, so that the data satisfies differential privacy before leaving the local device. In federated learning, each client can apply LDP processing to the data or gradients locally before sending them to the server for aggregation. Representative works include LDP-FL [6] etc.

3 Preliminaries

3.1 Federated Learning

Federated learning is a distributed machine learning framework that coordinates multiple participants to jointly train machine learning models [10]. It allows participants to collaboratively train a global model by transmitting only model parameters without sharing local data, thereby breaking the "data silo." Horizontal federated learning refers to situations where participating training parties have significant overlap in data features but minimal overlap in samples. For example, electricity companies in different regions have substantial overlap in the features of their electricity consumption data but minimal overlap in the samples themselves. Its training process is roughly as shown in Fig. 1.

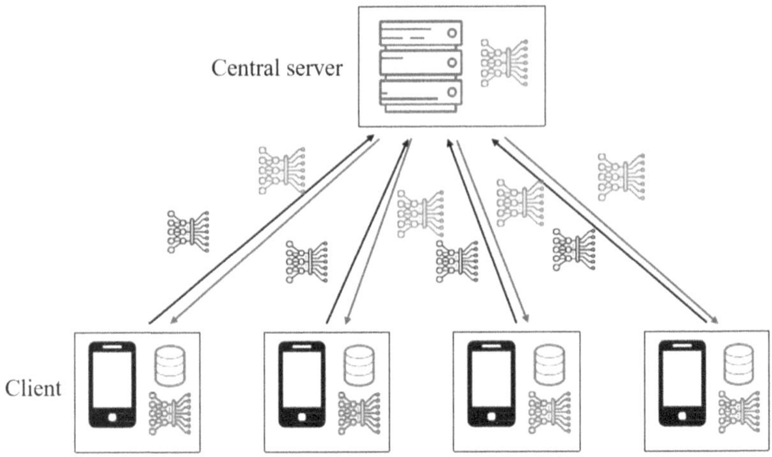

Fig. 1. Diagram of the horizontal federated learning training process

The federated averaging algorithm (FedAvg), proposed by Google, is the most classic and commonly used algorithm in the field of federated learning. Its core idea is to iteratively update the global model by aggregating the local model parameters from participating training nodes until the model reaches convergence conditions. As a groundbreaking work in federated learning, FedAvg introduces a simple and effective paradigm for distributed machine learning, laying the foundation for subsequent research. However, FedAvg exhibits poor performance when confronted with Non-IID data and privacy attacks. Most subsequent federated learning algorithms are based on FedAvg, with various improvements and innovations aimed at addressing its deficiencies.

3.2 Centralized Differential Privacy

In the centralized differential privacy model [7], the server first obtains the original data of all users to form a data set D. Then, the algorithm M, which satisfies differential privacy, is applied to the data set D for analysis and publication. Privacy protection is implemented on the server side, and users need to trust the server not to abuse data. The formal definition is described as follows:

Given two adjacent data sets D and D' (which differ by only one record), for a random algorithm M, its output is Ω. If the output of algorithm M on D and D' satisfies the following inequality, M provides $(\varepsilon, \bar{\delta})-$differential privacy protection:

$$\Pr[M(D) \in \Omega] \leq \exp(\varepsilon) \cdot \Pr\left[M\left(D'\right) \in \Omega\right] + \delta \tag{1}$$

where, the non-negative parameter ε is the privacy budget, which measures the degree of protection of individual privacy information by the algorithm M. The smaller the ε, the higher the degree of privacy protection, but the larger the output noise of the algorithm at this time. When ε approaches 0, the probability

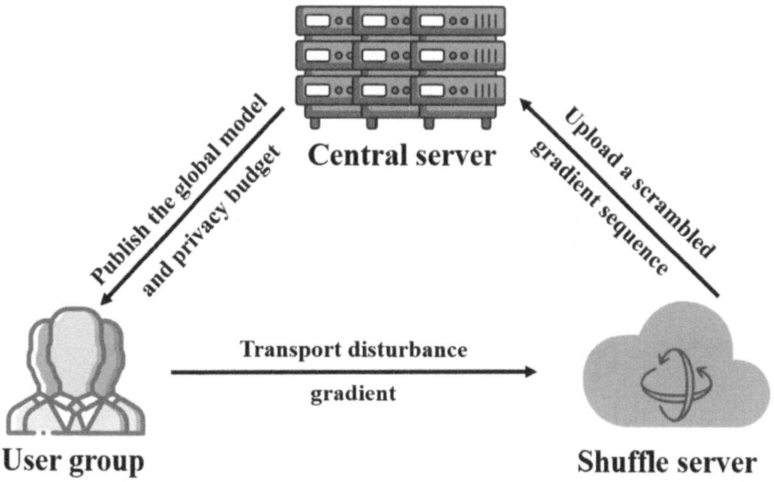

Fig. 2. SS-PFL system model

distribution of M output the same result on the adjacent data sets D and D' converges, and the algorithm M may leak less information. δ is also a non-negative parameter, indicating the probability that the difference between the output results of M on D and D' exceeds $\exp(\varepsilon)$, that is, the probability that the differential privacy protection is violated. Obviously, the smaller the δ, the higher the level of privacy protection.

4 System Model

This section introduces the entities involved in the SS-PFL algorithm, including clients, shuffle servers, and a central server, as well as the system model (Fig. 2).

- **User group:** Clients participating in collaborative training first obtain the latest global model parameters from the central server to initialize the local model. Subsequently, clients train a personalized model using their own local data to better adapt the model to their data. Clients then use the personalized model to optimize the training of the local model and obtain local model parameters. Finally, the client's local randomizer adds noise to the gradients based on the privacy budget and uploads the perturbed gradients to the shuffling server.
- **Shuffling Server:** It is a trusted third-party server independent of the central server, responsible for collecting the perturbed gradient sequences and scrambles their original correspondences. Subsequently, the shuffling server sends the shuffled gradient sequences to the central server for aggregation.
- **Central Server:** It is responsible for aggregating gradient information, updating the global model through aggregation algorithms, and broadcasting it to each client.

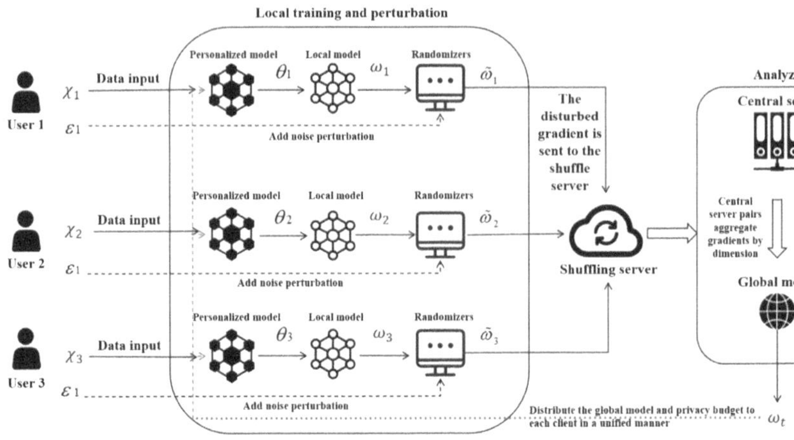

Fig. 3. Diagram of algorithm frame

5 Our Algorithm

The algorithm framework proposed in this paper is shown in Fig. 3, Algorithm 1, and Algorithm 2. The specific process of the SS-PFL algorithm is as follows:
System Initialization: The central server broadcasts the global model parameters ω_t and the privacy budget ε to all clients.

Local Update: Clients initialize their local model parameters based on the initial global model parameters ω_t broadcast by the central server. Subsequently, during the local model training phase, clients train personalized models $\tilde{\theta}_i\left(\omega_{i,r}^t\right)$ using their local datasets χ_i. Simultaneously, clients update their local models $\omega_{i,r}^t$ based on the computed personalized models $\tilde{\theta}_i\left(\omega_{i,r}^t\right)$ using the formula: $\omega_{i,r+1}^t = \omega_{i,r}^t - \eta\lambda\left(\omega_{i,r}^t - \tilde{\theta}_i\left(\omega_{i,r}^t\right)\right)$ until the local model $\omega_{i,R}^t$ is obtained. The central server randomly selects S clients to send their gradients to local randomizers M_i for perturbation. The randomizers then upload the perturbed gradient sequences $\{\tilde{\omega}_i\}_{i\in[n]}$ to the shuffle server, which generates permutations π_k for each dimension $k \in [d]$ of the gradients and randomly shuffles the gradient sequences based on these permutations.

Central Aggregation: The central server aggregates the gradients of the selected S clients for each dimension of the gradients using the formula: $\tilde{\omega} \leftarrow \frac{1}{S}\sum_{i\in S_t} \tilde{\omega}_{i,k}$, updates the global model, and broadcasts it to all clients: $\omega_{t+1} = (1-\beta)\omega_t + \beta\tilde{\omega}$.

Algorithm 1. Personalized federated learning algorithm based on shuffle differential privacy (SS-PFL)

Input: $T, R, S, A, n, B, \epsilon, |D|, w^0$
Output: Global model T
Initialize the global model w^0
for $t = 0, 1, 2, \ldots, T$ **do**
  Local update phase:
  Send global model w^t to all clients
  for $i = 1 \ldots N$ **do**
    $w_i^t = w^t$
    for $r = 1 \ldots R$ **do**
      Based on the size of new samples to minimize $|D|\ h_i^t(w, D_i^t)$, get personalized model approximation solution $tildew_i^t$
      $w_i^{t+1} = w_i^t - \eta(\nabla h_i^t(w_i^t, D_i^t) - \nabla h_i^t(\tilde{w}_i^t, D_i^t))$
    end for
    Sample the subset S_t of size S from the client collection
    $\tilde{R}_i \leftarrow \text{Randomize}(R_i)$
    The client uploads the perturbed gradient ω_i to the shuffle server
  end for
  Random shuffling stage:
  for $k \in [d]$ **do**
    $\pi_k$is generated based on $[d]$
    $\left\{(w_{i,\pi_k(k)}, k)\right\}_{i \in S_t} \leftarrow \text{Shuffle}(k, \{D_{i,k}\}_{i \in S_t})$
  end for
  Central aggregation:
  for $k \in [d]$ **do**
    $\omega_k \leftarrow \frac{1}{|S_t|} \sum_{i \in S_t} \omega_{i,\pi_k(k)}$
  end for
  $w^{t+1} = (1 - B)w^t + B \sum_{k=1}^{d} \omega_k$
end for
Return Global model $T = w^{T+1}$

Algorithm 2. The perturbation algorithm (Randomize)

Input: $\left\{(\omega_i, \varepsilon_i)\right\}_{i \in [n]}$
Output: The gradient after the disturbance $\tilde{\omega}_i$
$\tilde{\omega}_i \leftarrow \mathbf{Lap}\,(0, (d\Delta f)/\varepsilon_i)$
$I_b \leftarrow \left\{k \mid k \in \max\left(|\tilde{\omega}_{i,k}|_{k \in [d]}\right)\right\}^b$
for $k \notin I_b$ **do**
  $\tilde{\omega}_{i,k} \leftarrow \mathbf{Lap}\,(0, (d\Delta f)/\varepsilon_i)$
end for
Return local model W_k

6 Experimental Analysis

This subsection primarily conducts experimental verification of the personalized federated learning algorithm based on shuffled differential privacy. Under the

Non-IID data setting, the MNIST dataset is selected to perform a comparative analysis of the algorithm's ability to resist reconstruction attacks and the accuracy of the model. The hardware and software configuration of this experiment is shown in Table 1.

Table 1. Hardware and software configuration of the experiment

Environment	Configuration	Environment	Configuration
CPU model	I7 13700 k	Linux	Ubuntu22.04.1 LTS
Memory	32 GB	VGA Drivers	551.86.00
GPU	RTX 3090	CUDA	1.7
Video memory	24 GB	Python	3.12.2
Hard disk	2TB	Tensorflow	2.5

6.1 Experimental Data Sets and Partitioning Methods

MNIST is a handwritten digit dataset that contains 70,000 grayscale images of the same size, 28×28 pixels, with 60,000 images for training and 10,000 images for testing. The dataset includes handwritten digits from 0 to 9, totaling 10 categories.

To simulate the client-side Non-IID data environment, this paper uses Dirichlet distribution to split the raw data set and send the uneven data to all clients. A Dirichlet distribution is a joint probability distribution that defines multiple random variables on the interval [0,1], denoted as $Dir(\alpha)$. The specific steps to split the data set are as follows:

(1) Set the parameter α to simulate different degrees of data heterogeneity by adjusting the parameter α, where the smaller the value of α, the more unbalanced the generated data segmentation, that is, the number of samples of some categories in some clients may be much higher than others.
(2) For each client participating in federated training, randomly generate a K-dimensional vector $p = (p_1, p_2, ...p_k)$ from the Dirichlet distribution, where K represents the number of categories in the dataset (for MNIST, $K = 10$). The vector $p$represents the sample proportion of each category in that client.
(3) For each sample, according to its category label and vector p, decide whether it is assigned to the current client. The process can be viewed as a polynomial distribution sampling process, where the vector $p$determines the sampling probability for each class.
(4) Repeat steps 2 and 3 until all client data is allocated. The random sampling process is denoated as $p \sim dir(\alpha)$.

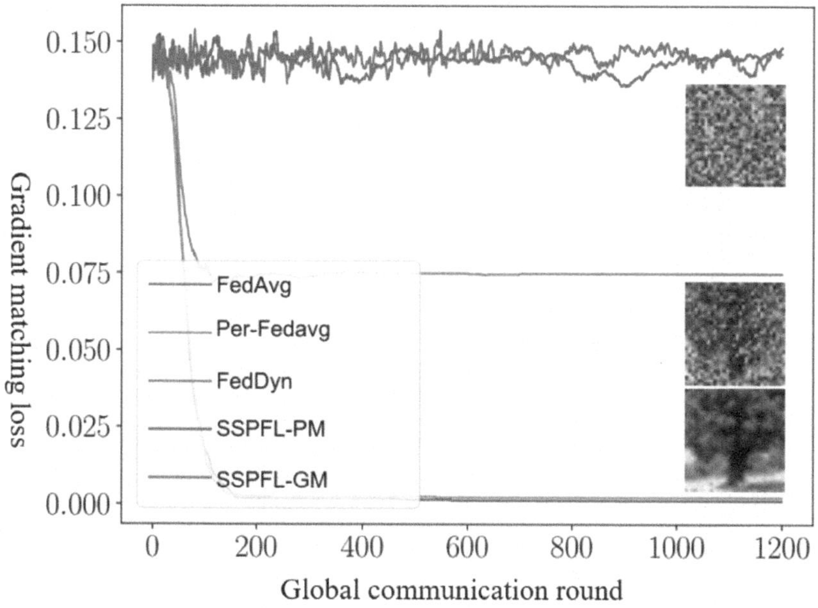

Fig. 4. SS-PFL resistance to reconstruction attack performance.

6.2 Algorithm Performance Analysis

To verify that SS-PFL can resist data reconstruction attacks, the DLG reconstruction attack was applied to different algorithms under the same settings based on the privacy budget. The vertical axis represents the gradient matching loss mentioned earlier, and the horizontal axis represents the number of iterations. As can be seen from Fig. 4, on the MNIST dataset, under the same noise intensity, the DLG attack still struggles to recover recognizable original data from the gradient information of SS-PFL, while other methods such as FedAvg and Per-FedAvg cannot resist the DLG reconstruction attack. This fully demonstrates the advantage of SS-PFL in resisting data reconstruction attacks - even if a malicious attacker obtains the gradient information of a certain client, they cannot recover the training data from the noise.

To verify the improvement of SS-PFL on model performance under Non-IID conditions, we first compared algorithms such as FedAvg, Per-FedAvg, and FedDyn using the same hyperparameters. Figures 5, 6 show that the personalized model of SS-PFL excels in test accuracy and training loss, surpassing its global model, FedDyn, and Per-FedAvg [5] by 0.015, 0.005, and 0.014 respectively. Although the global model of SS-PFL performs less favorably than other algorithms on these two metrics, its personalized model demonstrates advantages, achieving the highest test accuracy and the lowest training loss.

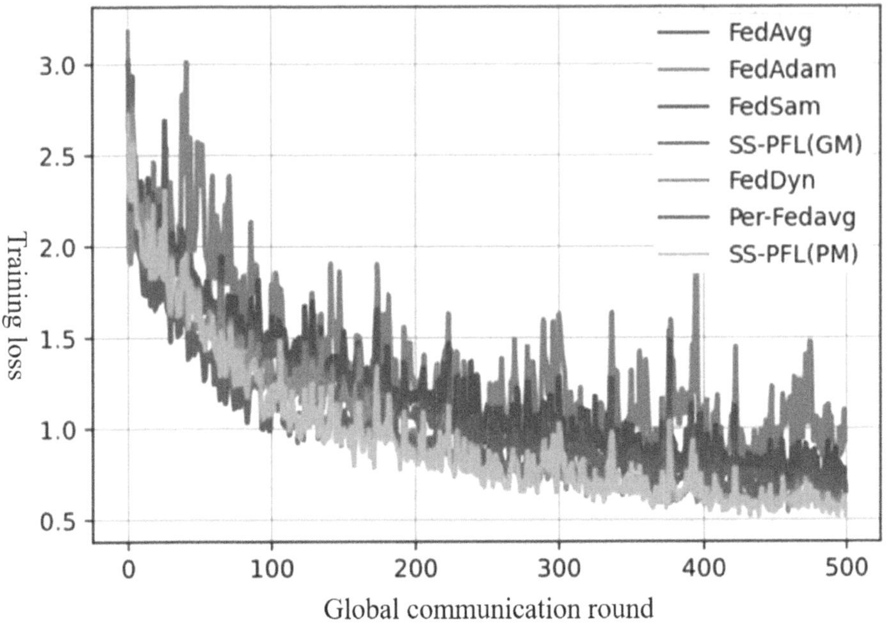

Fig. 5. Loss performance comparison between SS-PFL and other algorithms on MNIST.

However, due to the variations in algorithm performance resulting from different hyperparameter settings, a grid search was conducted across a wide range of hyperparameters to identify the fine-tuning parameter combination that yields the highest accuracy for each algorithm. The comparison results based on fine-tuning parameters are shown in Table 2: on the MNIST dataset, the personalized model of SS-PFL performs best under all settings. Furthermore, the introduction of a shuffle server ensures privacy security during the aggregation process but also incurs additional communication costs. As shown in Table 3, the convergence speed of SS-PFL on the MNIST dataset is within an acceptable range.

Table 2. Algorithm comparison under fine tuning parameters

Method	MNIST		
	λ	$\eta(\hat{\alpha}, \hat{\beta})$	Accuracy(%)
FedAvg		0.02	68.02 ± 0.03
PerFedAvg		0.03,0.003	69.91 ± 0.06
SS-PFL(GM)	15	0.01	70.93 ± 0.06
SS-PFL(PM)	15	0.01	71.91 ± 0.02

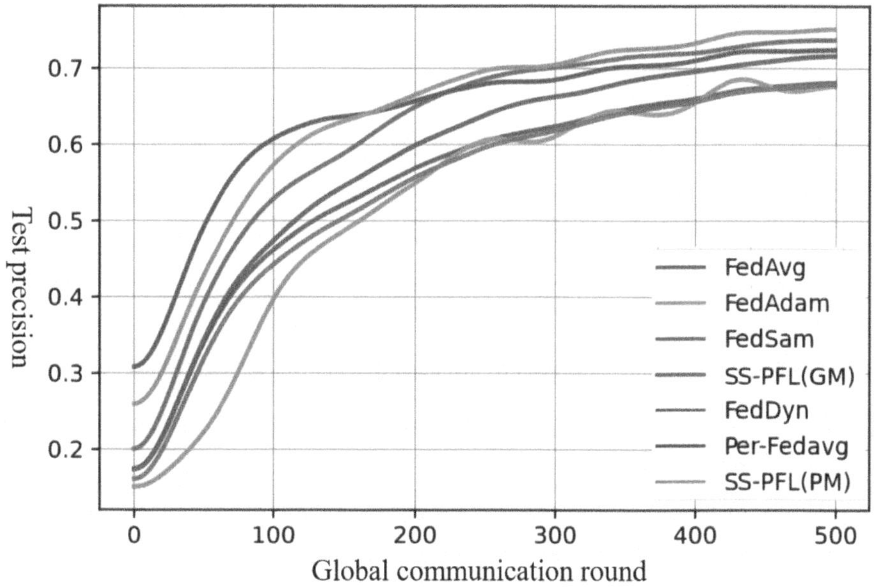

Fig. 6. Comparison of test accuracy between SS-PFL and other algorithms on MNIST.

Table 3. Communication overhead of different algorithms

Method	MNIST	
	Round	Accuracy(%)
FedAvg	356	65
PerFedAvg	327	65
SS-PFL(GM)	452	65
SS-PFL(PM)	423	65

7 Conclusion

In order to ensure the performance of federated learning models in non-independent and identically distributed (Non-IID) environments while effectively protecting the data privacy of training parties, this chapter innovatively proposes a personalized federated learning framework based on shuffled differential privacy, namely SS-PFL. The framework initially employs a bi-level optimization strategy to train personalized models locally on the client side, thereby optimizing and enhancing the performance of local models. Furthermore, to further enhance privacy protection, the framework introduces a shuffle server to ensure the secure aggregation of gradient parameters under unified privacy budget constraints. Experimental results demonstrate that, compared to existing privacy-preserving federated learning methods, our proposed algorithm not only achieves higher classification accuracy but also effectively resists reconstruction attacks, showcasing its exceptional performance and robust privacy protection capabilities.

References

1. Bonawitz, K., et al.: Towards federated learning at scale: system design. Proc. Mach. Learn. Syst. **1**, 374–388 (2019)
2. Cai, L., Chen, N., Cao, Y., He, J., Li, Y.: FedCE: personalized federated learning method based on clustering ensembles. In: Proceedings of the 31st ACM International Conference on Multimedia, pp. 1625–1633 (2023)
3. Chan, Y.H., Ngai, E.C.: FedHE: heterogeneous models and communication-efficient federated learning. In: 2021 17th International Conference on Mobility, Sensing and Networking (MSN), pp. 207–214. IEEE (2021)
4. Cheng, K., et al.: SecureBoost: a lossless federated learning framework. IEEE Intell. Syst. **36**(6), 87–98 (2021)
5. Fallah, A., Mokhtari, A., Ozdaglar, A.: Personalized federated learning with theoretical guarantees: a model-agnostic meta-learning approach. In: Advances in Neural Information Processing Systems, vol. 33, pp. 3557–3568 (2020)
6. Jabid, T., Kabir, M.H., Chae, O.: Local directional pattern (LDP)–a robust image descriptor for object recognition. In: 2010 7th IEEE International Conference on Advanced Video and Signal Based Surveillance, pp. 482–487. IEEE (2010)
7. Liu, Z., Hu, C., Xia, H., Xiang, T., Wang, B., Chen, J.: SPDTS: a differential privacy-based blockchain scheme for secure power data trading. IEEE Trans. Netw. Serv. Manage. **19**(4), 5196–5207 (2022). https://doi.org/10.1109/TNSM.2022.3181814
8. Ma, J., Naas, S.A., Sigg, S., Lyu, X.: Privacy-preserving federated learning based on multi-key homomorphic encryption. Int. J. Intell. Syst. **37**(9), 5880–5901 (2022)
9. McMahan, B., Moore, E., Ramage, D., Hampson, S., y Arcas, B.A.: Communication-efficient learning of deep networks from decentralized data. In: Artificial Intelligence and Statistics, pp. 1273–1282. PMLR (2017)
10. Wang, B., Hu, C., Liu, Z.: A secure aggregation scheme for model update in federated learning. In: Wang, L., Segal, M., Chen, J., Qiu, T. (eds.) Wireless Algorithms, Systems, and Applications, pp. 500–512. Springer Nature Switzerland, Cham (2022)
11. Wang, P., Liu, Y., Li, Z., Li, R.: An LDP compatible sketch for securely approximating set intersection cardinalities. Proc. ACM Manage. Data **2**(1), 1–27 (2024)
12. Zhang, P., Chen, N., Li, S., Choo, K., Jiang, C., Wu, S.: Multi-domain virtual network embedding algorithm based on horizontal federated learning. IEEE Trans. Inf. Forensics Secur. **18**, 3363–3375 (2023). https://doi.org/10.1109/TIFS.2023.3279587
13. Zhao, Y., Chen, J.: A survey on differential privacy for unstructured data content. ACM Comput. Surv. (CSUR) **54**(10), 1–28 (2022)

Research and Design of Tomato Detection Accelerator Based on FPGA

Bin Yu(✉), Wanyu Lu, Shangwen Sui, Zhiwei Liu, and Ao Li

School of Computer Science and Technology, Harbin University of Science and Technology, No. 52 Xuefu Road, Harbin 150080, Heilongjiang Province, China
yubin@hrbsut.edu.cn

Abstract. In order to solve the problem of real-time detection of tomato quantity, location and maturity, a real-time tomato detection scheme based on FPGA was designed. The scheme uses the quantizer and compiler in the Vitis-AI tool to quantize and compile the YOLOv5s algorithmic model respectively, obtains the optimized model and deploys it to the FPGA accelerator platform, and then uses the ZCU104 FPGA accelerator to accelerate the inference of the YOLOv5s algorithmic model, so as to be able to realize the real-time detection task of tomato on the hardware platform. The experimental results show that the recognition effect of the YOLOv5s network in this paper has a certain improvement over the comparison YOLOv5s network in terms of precision rate, recall rate, and mean average precision rate.

Keywords: Object Detection · YOLOv5s · FPGA · Vitis-AI

1 Introduction

Tomato is one of the three major vegetables in the world, playing a crucial role in the global vegetable trade. In the process of tomato growth, it is necessary to dynamically and accurately detect the number, position and maturity of tomatoes. When estimating the yield of fruits at different stages of ripeness, challenges such as leaf occlusion, dense growth, and difficulty in distinguishing ripeness pose significant issues [1]. At the same time, traditional tomato detection methods rely on manual harvesting and observation, which faces challenges in estimating yield and ensuring accuracy. Fruit detection has profited from the application of computer vision in recent years, an affordable and effective method [2]. As an important research field of computer vision, target detection technology also plays an important role in tomato detection. Among various object detection algorithms, YOLO (You Only Look Once) stands out for its good detection results and lower complexity, which presents the best trade-off between accuracy and execution time [3].

X. Li et al. (Eds.): 6GN 2024, LNICST 615, pp. 191–198, 2025.
https://doi.org/10.1007/978-3-031-93193-2_16

Due to the limitation of computing power, traditional object detection systems have some problems, such as high transmission latency, complex system and low reliability. To address these problems, it is of great significance to deploy object detection algorithms on edge computing platforms [4]. Edge computing platforms usually include CPUs, GPUs, ASICs, and FPGAs. CPUs have the advantage of high performance and versatility, but they cannot provide sufficient computational power for real-time object detection. GPUs offer design flexibility but consume a considerable amount of power, making them not suitable for many IoT or edge computing applications [5]. ASICs consume less power than GPU but sacrifice flexibility [6]. Compared to CPUs, GPUs and ASICs, FPGAs have the characteristics of high performance, low power consumption, programmable and parallel computing mode, which makes them play a stable and reliable accelerating role in tomato detection. Therefore, this paper designs a method of an FPGA based on the Vitis-AI tool accelerated implementation of the YOLOv5s algorithm modeling, which enables real-time target detection for tomatoes.

2 YOLO Algorithm Model Introduction and Dataset Construction

2.1 YOLO Algorithm Model Introduction

YOLO (You Only Look Once) algorithms are a series of deep learning algorithms used for real-time object detection. The main idea of this series of algorithms is to transform the object detection problem into a single regression problem. They extract features from the input image and then directly locate and classify the target on the extracted features, thus achieving fast and accurate object detection [7]. At present, YOLO algorithms have developed to the latest YOLOv10, and each version has been improved and optimized on the basis of the previous version to improve the detection accuracy and speed.

YOLOv5 is the fifth version of the YOLO algorithms. It combines the features of the YOLOv1-YOLOv4 versions and has been improved in terms of detection speed and accuracy, and it is widely used in the field of object detection with its excellent performance [8]. The YOLOv5 network framework is composed of input, backbone, neck and head, with the overall framework shown in Fig. 1. Firstly, images are processed through the input layer and sent to the backbone for feature extraction. Secondly, in the backbone part, YOLOv5 adopts CSPDarknet53 as the backbone network, which combines the design ideas of CSP (Cross Stage Partial) and Darknet, enabling the algorithm to extract features from the input images faster for subsequent object detection tasks. Thirdly, the neck part, which is between the backbone and the head, is used to integrate feature maps at different scales. Finally, the head structure contains multiple levels of prediction, which is responsible for predicting targets of different sizes. YOLOv5 algorithm includes four versions: YOLOv5x, YOLOv5l, YOLOv5m, and YOLOv5s. Among them, the YOLOv5s network model is the network with the smallest depth and the smallest feature map width in the series of YOLOv5, and the three models of YOLOv5m, YOLOv5l and YOLOv5x are the products of continuous deepening and widening on the basis of YOLOv5s [9]. Taking into consideration factors such as speed and accuracy, this study adopts YOLOv5s as the design model.

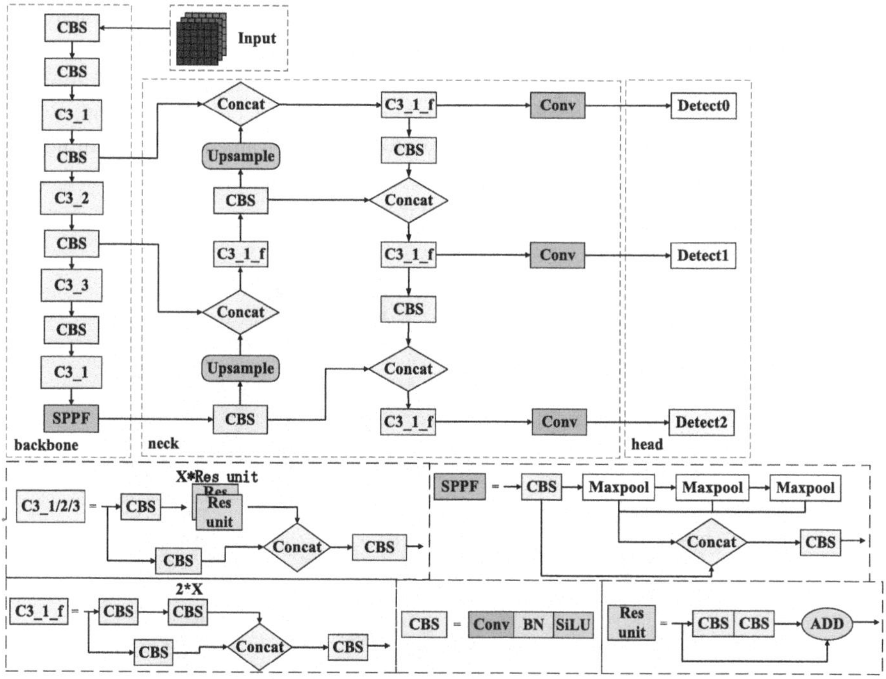

Fig. 1. Overall Structure of YOLOv5

2.2 Improved YOLOv5s Model

Based on the analysis of the YOLO model in Subsect. 2.1, this paper proposes an improved YOLOv5s model. In this model, three feature maps of different scales, generated by the original networks at layers 4, 6, and 9, are input into the neck layer. Additionally, the feature map from layer 2 is also introduced into the neck layer. Thus, four feature maps from layers 2, 4, 6, and 9 are jointly input into the neck layer. Next, the feature maps from the backbone layer are fused and processed by the FPN and PAN structures. Finally, the Detect layer generates feature maps at four different scales from layers 21, 24, 27, and 30.

2.3 Dataset Construction

Datasets are the basis for training machine learning models. Models learn patterns, features, and relationships of the input data through datasets, so as to carry out prediction, classification, clustering and other tasks. Datasets are also used to evaluate and validate the performance of trained models. Generally, datasets are divided into training set, validation set, and test set. In this paper, about 2700 images are collected independently to construct the dataset, and some of the dataset images are shown in Fig. 2. Among them, both training set and validation set are used in the training process. The training set is used to train the machine learning model, enabling the model to learn patterns and features from the data, adjust its parameters, and minimize prediction errors. The function of validation set is to verify the performance of the model under different hyper-parameter settings (e.g., learning rate, regularization parameter, etc.), so that people can select the best model configuration. The purpose of test set is to evaluate the model's performance metrics such as accuracy, precision, recall, and other indicators, which reflect the generalization ability and accuracy of the model trained on the training set. Because the model requires enough data to learn effectively, the training set usually accounts for a large percentage of the total dataset.

Fig. 2. Some Images from the Dataset

3 Model Conversion and Deployment

Vitis-AI is a tool and framework launched by Xilinx for accelerating AI inference, which focuses on accelerating the inference process of deep learning models with Xilinx hardware accelerators (e.g., FPGAs, ACAPs). It is composed of optimized IP, tools, libraries, models, and example designs [10]. Throughout the accelerated inference process, Vitis-AI plays the role of compiler and back-end, receiving the network parameters IR (Intermediate Representation) trained by the front-end DNN (Deep Neural Network) framework, compiling it after optimization and passing it to the back-end. The back-end DNNDK (Deep Neural Network Development Kit) provides driver and API for the Edge terminal, as well as tool library for debugging and running. In this paper, we use the Vitis-AI tool to quantify and compile the YOLOv5s algorithmic model and deploy it to an FPGA accelerator so that the accelerator can accelerate the input image or video stream in real-time.

The design process of Vitis-AI covers three main components: model construction, software construction, and hardware construction, as shown in Fig. 3. Model and software construction occur on the software side, while hardware construction takes place on the hardware side. Through the cooperation of software and hardware, Vitis-AI completes

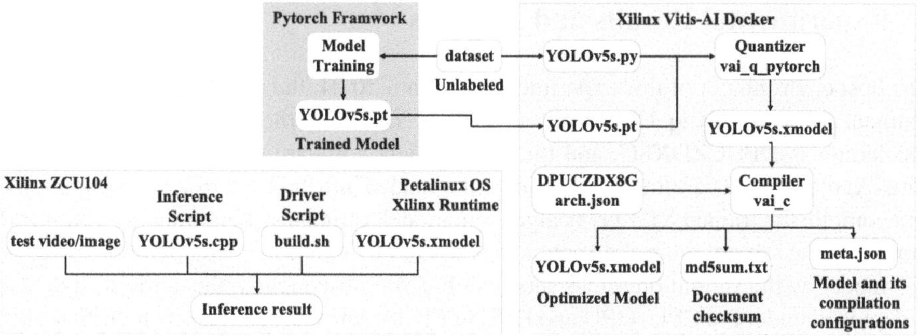

Fig. 3. Overall Deployment Flowchart

end-to-end design and deployment tasks, which makes AI model efficiently deployed and run on Xilinx FPGA accelerators.

Specifically, the YOLOv5s algorithm model is constructed using Pytorch framework, which takes floating-point form and dataset as training inputs, and the YOLOv5s floating-point model is obtained after training is completed. Then, in the build software section, we use the Vitis-AI tool library to optimize and transform the trained model. As shown in Fig. 3, we use the AI quantizer to convert the 32-bit floating-point data (FP32) of the YOLOv5s model to an 8-bit integer (INT8) in the Xilinx Vitis-AI Docker environment, which reduces the size and computation of the model, and improves the inference speed and efficiency of the model on FPGA accelerator. After quantization, the AI compiler is used to compile the model, so that the model trained in Pytorch framework can be compiled into the model accelerated by DPU (Deep Learning Processing Unit). Whereas DPUs are programmable engines dedicated to convolutional neural network (CNN) acceleration, they have a specialized instruction set that enables DPUs to efficiently provide computing tasks for the convolutional part of many deep neural networks. People integrate DPU IP as a module into the programmable logic (PL) of FPGA hardware accelerator and connect it to the processing system (PS) through AXI bus.

Because Vitis-AI provides a special runtime library and API, it can communicate with hardware accelerator by using C++ or Python language on the software side. Through these APIs, developers can manage the startup and operation of the hardware accelerator on the PS, as well as implement efficient management of data flow and task scheduling in the PL. Therefore, in the part of hardware construction, this paper chooses Xilinx ZCU104 FPGA development board and builds petalinux environment on this hardware platform to provide an embedded Linux system for Xilinx's Zynq MPSoC platform. Then, the optimized model is deployed to the DPU (PL side) of this hardware platform, so that the Vitis-AI runtime (Xilinx Runtime) environment can be invoked through C+ + or Python language, which enables the PS side to load and initialize the FPGA accelerator on the software interface. Finally, during the inference process of YOLOv5s algorithm model, the input data is received through the network interface for preprocessing and data preparation, so it can be transmitted to the PL side to execute the calculation tasks and generate the output results, and then the output results are transmitted back to the PS side for the final output results.

4 Experimental Results and Analysis

The host environment of this experiment is Ubuntu 20.04, the development board environment is Xilinx Zynq UltraScale + MPSoC ZCU104, the end-side deep learning accelerator is DPUCZDX8G, and the development software mainly uses Docker and Vitis-AI of Xilinx. The Vitis-AI environment is loaded into the Ubuntu system to quantify and compile the trained YOLOv5s algorithm model, ultimately obtaining the optimized model. At the same time, we also obtain the result of validating the recognition effect of this model by the validation set, as shown in Fig. 4. Subsequently, this optimized model is deployed under the CPU, GPU and DPU of FPGA development board on the host side respectively.

Fig. 4. Tomato Recognition Results

Figure 4 shows the recognition effect of YOLOv5s network model in this paper, where mature, semi-mature, and immature represent the maturity of tomato, respectively. In Table 1, we compare the detection effects of the original YOLOv5s network and its improved YOLOv5s network based on the dataset of Reference [11], as well as the original YOLOv5s version 5.0 and 6.0 and improved YOLOv5s model networks based on the dataset of this paper. The evaluation metrics in the table include precision, which measures the accuracy of positive predictions; recall, which measures the recognition rate of positive samples; and mAP(mean Average Precision), which synthesizes these metrics to evaluate the overall performance of the model in multi-category detection. Both this paper and Reference [11] aim to implement a real-time detection scheme based on FPGA. However, the improved YOLOv5s network in Reference [11] shows a decrease in recall and mAP metrics compared to the original YOLOv5s network, which will lead to an increase in the number of recognition errors and omissions of the target to be detected. Compared to the original YOLOv5s network, the improved YOLOv5s network proposed in this paper maintains both precision and recall, while effectively improving the mAP@0.5–0.95 accuracy. And compared with the improved YOLOv5s network in Reference [11], the effect of this paper's improved network is more significant, and compared with the original YOLOv5s network, this paper's improved network is more accurate, comprehensive, and stable for the prediction of tomato maturity.

Table 1. Recognition results compared with other YOLOv5s network

Network	Precision	Recall	mAP@0.5	mAP@0.5–0.95
Original YOLOv5s-5.0 network in Reference [11]	79.4	79.1	41.7	83.5
Improved YOLOv5s network in Reference [11]	79.5	76.9	40.3	81.7
Original YOLOv5s-5.0 network in this paper	98.9	93.6	96.9	60.6
Original YOLOv5s-6.0 network in this paper	97.4	92.2	96.0	71.8
Improved YOLOv5s-6.0 network in this paper	96.7	92.0	95.9	76.9

5 Conclusion

This study uses the YOLOv5s algorithm model to design and implement a tomato detection accelerator based on FPGA. The YOLOv5s algorithm model is quantized and compiled using the Vitis-AI tool, and accelerated inference on the FPGA hardware accelerator, which solves the problem of real-time detection of tomato as well as the problem of ripeness recognition of tomato. The experimental results demonstrate that the YOLOv5s network model proposed in this study achieves superior performance, particularly on FPGA hardware accelerator.

Acknowledgement. This work was supported by the National Key Research and Development Program of China (Grant No. 2022YFD2000500).

References

1. Wang, A., et al.: NVW-YOLOv8s: an improved YOLOv8s network for real-time detection and segmentation of tomato fruits at different ripeness stages. Comput. Electron. Agric. **219**, 1–10 (2024)
2. Yang, R., He, Y., Hu, Z., Gao, R., Yang, H.: CA-YOLOv5: a YOLO model for apple detection in the natural environment. Syst. Sci. Control Eng. **12**(1), 1–10 (2024)
3. Pestana, D., et al.: A full featured configurable accelerator for object detection with YOLO. IEEE Access **9**, 75864–75877 (2021)
4. Cao, J., Yang, Z., Lu, J., Lai, J.: A high-performance YOLOV5 accelerator for object detection with near sensor intelligence. In: 2023 IEEE 15th International Conference on ASIC (ASICON), pp. 1–4. IEEE, Nanjing, China (2023)

5. Kim, M., Oh, K., Cho, Y., Seo, H., Nguyen, X.T., Lee, H.-J.: A low-latency FPGA accelerator for YOLOv3-tiny with flexible Layerwise mapping and dataflow. IEEE Trans. Circuits Syst. I Regul. Pap. **71**(3), 1158–1171 (2024)
6. Farrukh, F.U.D., et al.: Power efficient tiny YOLO CNN using reduced hardware resources based on booth multiplier and WALLACE tree adders. IEEE Open J. Circ. Syst. **1**, 76–87 (2020)
7. Wang, X., et al.: Lightweight SM-YOLOv5 tomato fruit detection algorithm for plant factory. Sensors **23**, 1–19 (2023)
8. Wang, B., Yan, Y., Lan, Y., Wang, M., Bian, Z.: Accurate detection and precision spraying of corn and weeds using the improved YOLOv5 model. IEEE Access **11**, 29868–29882 (2023)
9. Chen, T., Ding, Z., Li, B.: Elderly fall detection based on improved YOLOv5s network. IEEE Access **10**, 91273–91282 (2022)
10. Wang, J., Gu, S.: FPGA implementation of object detection accelerator based on Vitis-AI. In: 11th International Conference on Information Science and Technology (ICIST), pp. 571–577. IEEE, Chengdu, China (2021)
11. Liu, Q., Wang, L., Zhou, W.: Design of a YOLOv5s network efficient convolution accelerator powered by FPGA. Telecommun. Eng. **64**(3), 366–375 (2024)

Research on the Role of Fiscal Deficit in the Impact of Digital Inclusive Finance on Economic Growth

Xiaofeng Zhang and Fangfang Ji(✉)

School of Public Finance and Administration, Harbin University of Commerce, Harbin 150028, Heilongjiang, China
jiffang@163.com

Abstract. The fiscal deficit, as one of the means to regulate macroeconomic operations, plays an important role in promoting digital inclusive finance and economic growth. Based on panel data from 31 provinces from 2011 to 2022, this study examines the mechanism of fiscal deficits in the impact of digital inclusive finance on economic growth. The research results indicate that fiscal deficits have a positive moderating effect and a single threshold effect on the promotion of economic growth through digital inclusive finance. The heterogeneity of the moderating effect is strong in the eastern and western regions, while there is no effect in the central region; When the proportion of livelihood fiscal expenditure is low, it is strong, but when the proportion is high, it has no effect. The single threshold effect is manifested as a non-linear characteristic of the promotion effect of digital inclusive finance on economic growth showing a first strong and then weak trend as the scale of fiscal deficits expands. Based on this, countermeasures and suggestions are proposed to implement differentiated fiscal deficits, improve the efficiency of fund utilization, and set reasonable upper limit standards for deficits, in order to fully leverage the macroeconomic regulation role of fiscal deficits.

Keywords: Digital Inclusive Finance · Fiscal Deficit · Economic Growth

1 Introduction

With the widespread application of the Internet, artificial intelligence, and 5G technology, digital inclusive finance has emerged. It has the characteristics of digitalization and inclusiveness, creating a good financial environment for economic growth. Specifically, digital inclusive finance has broken down traditional financial service barriers and promoted economic growth in areas such as consumer entrepreneurship, rural poverty reduction, technology spillovers, and enterprise innovation. As one of the positive fiscal policy tools, fiscal deficit plays an important role in stimulating consumption, expanding demand, and stimulating employment, with the ultimate goal of promoting economic growth [1].The 2024 Government Work Report shows that the annual deficit rate is 3%, and the deficit size is 4.06 trillion yuan. This year, 1 trillion yuan of ultra long-term special treasury bond will be issued. Overall, China's deficit scale is moderate and has not

X. Li et al. (Eds.): 6GN 2024, LNICST 615, pp. 199–208, 2025.
https://doi.org/10.1007/978-3-031-93193-2_17

exceeded the 3% warning line, which is significantly lower than that of similar countries, and there is still some room for expansion [2]. It can be inferred that both digital inclusive finance and fiscal deficits contribute to economic growth. So, what role does fiscal deficit play in the impact of digital inclusive finance on economic growth? Based on this, this article conducts empirical research, hoping to provide new theoretical basis and analytical perspectives for China's economic development. The possible contributions of this article are: firstly, it demonstrates that fiscal deficits have a positive regulatory effect in promoting economic growth through digital inclusive finance; Secondly, a heterogeneity analysis of the proportion of regional and livelihood fiscal expenditures was conducted on the regulatory effect of fiscal deficits; Thirdly, it was found that fiscal deficits have a single threshold effect on promoting economic growth through digital inclusive finance.

2 Related Literature and Research Hypotheses

2.1 The Moderating Effect of Fiscal Deficit on the Promotion of Economic Growth Through Digital Inclusive Finance

In recent years, digital inclusive finance has mainly driven economic growth at four levels: household consumption and entrepreneurship, rural poverty reduction, technology spillovers, and enterprise innovation. From existing research literature, there have been many studies on fiscal deficits, with domestic and foreign scholars mainly focusing on economic effects, influencing factors, fiscal sustainability, and other aspects. Fiscal deficit refers to the ratio of the difference between general public budget expenditure and revenue to GDP, which is directly caused by the expenditure demand to address public risks exceeding fiscal revenue within a given budget cycle. The essence of digital inclusive finance is profit seeking behavior, which requires the government to invest a certain amount of fiscal funds to upgrade financial infrastructure, fill the gaps in financial development, improve digital financial facilities, and stimulate its potential for economic growth. Generally speaking, the government obtains deficit funds by issuing public debt, thereby providing financial support for financial development. Therefore, within a reasonable range, the higher the fiscal deficit, the more sufficient the funds, and the faster the development of digital inclusive finance, the stronger its promoting effect on economic growth. Therefore, the hypothesis is proposed:

H1: The fiscal deficit has a positive regulatory effect on the promotion of economic growth through digital inclusive finance.

2.2 Heterogeneity Analysis of the Regulatory Effect of Fiscal Deficits on Digital Inclusive Finance in Promoting Economic Growth

The regulatory effect of fiscal deficits may inevitably be influenced by the following factors. One is the influence of geographical regions. China has a vast territory, and uneven economic development between regions is a major characteristic of China's economic development [3]. Differences in geographical environment and resource endowments often lead to significant differences in the external environment of financial development in different regions. For example, the eastern region is close to the coast, with a superior geographical location, abundant financial resources and complete financial infrastructure construction. The central region is densely populated, and industries mainly rely on the development of agriculture and mineral resources. The level of technological innovation and development is relatively limited. The western region is vast, with low population density and relatively remote geographical location, lacking access to high-quality financial resources. The second is the impact of the proportion of fiscal expenditure on people's livelihoods. It is a fundamental expenditure to promote stable and orderly economic development, mainly including expenditures in education, healthcare, public safety, social security and employment, and environmental protection [4]. Due to the important position of livelihood fiscal expenditure in the entire expenditure structure, it will to some extent squeeze out constructive fiscal expenditure, affecting the use of fiscal deficit funds. When the proportion is low, fiscal deficit funds are usually used to develop high-end industries; When the proportion is high, most of the fiscal deficit funds are used to solve basic livelihood problems, and the investment in financial development will decrease. Therefore, the hypothesis is proposed:

H2: The regulatory effect of fiscal deficits exhibits heterogeneity in different regions and the proportion of people's livelihood fiscal expenditures.

2.3 The Threshold Effect of Fiscal Deficit on Promoting Economic Growth Through Digital Inclusive Finance

The modern theoretical origin of fiscal deficits can be traced back to Keynes (1883)'s "functional fiscal theory", who advocated that finance should be considered as a total demand and included in the scope of macroeconomic policies, and proposed a "discretionary" fiscal policy. Starting from this policy, during economic downturns, the government will spare no effort in arranging deficits to increase spending and expand overall social demand. Nowadays, fiscal deficits have become a universal phenomenon worldwide, characterized by normalization and long-term effects. Fiscal deficits can not only solve the problem of insufficient budget funds, but also regulate the macroeconomy. However, fiscal deficit is a double-edged sword that requires the government to choose an appropriate scale of deficit based on the actual situation, in order to avoid excessive fiscal deficit and increase the burden on local governments. Specifically, within a reasonable range, fiscal deficits can stimulate consumption and investment, expand total social demand, increase household employment [5]., and be beneficial for regional financial infrastructure construction, financial industry development, and technological innovation. However, when the fiscal deficit is too large, it may have negative effects such as inflation, currency depreciation, and misallocation of scarce resources, which can disrupt the stability of financial markets. At the same time, a large fiscal deficit will also increase the pressure on local governments to repay debts, forcing them to invest fiscal funds in areas with high short-term returns and quick results, thereby reducing investment in financial development. Therefore, the hypothesis is proposed:

H3: The threshold effect of fiscal deficit in promoting economic growth through digital inclusive finance has non-linear characteristics.

3 Research Design

3.1 Data Sources and Variable Selection

The index related to digital inclusive finance comes from the report of Peking University Digital Inclusive Finance Index [6], and other data comes from the National Bureau of Statistics and EPS database.

The dependent variable is economic growth (LnPGDP). Measure by taking the natural logarithm of per capita GDP.

Core explanatory variables: Digital Inclusive Finance Index (Difi) and its coverage breadth (Cob), depth of use (USD), and degree of digitization (Dil) sub indices.

Adjusting variable: Fiscal deficit (Fde). Measure by the proportion of the difference between general public budget expenditure and income to GDP.

Control variables: To control for the interference of other indicators on the empirical results during the research process, this article selected government support level (Gov), foreign trade dependence (Open), industrial structure adjustment (Is), urban resident income (Lnicm), and labor factor (Upd) as control variables.

3.2 Model Settings

Firstly, through the Hausman test, it was found that the fixed effects model is suitable for this study, so the benchmark regression model is set as follows.

$$\mathrm{LnPgdp}_{i,t} = \alpha_1 + \beta_1 \mathrm{Difi}_{i,t} + \theta \mathrm{Control}_{i,t} + \mu_i + \tau_i + \varepsilon_{i,t} \tag{1}$$

Secondly, based on Eq. (1), the interaction term $\mathrm{Difi}_{i,\,t} * \mathrm{Fde}_{i,\,t}$ between digital inclusive finance and fiscal deficit as the moderating variable and conducts decentralization treatment to construct a moderating effect model as shown in Eq. (2) [7]:

$$\mathrm{LnPgdp}_{i,t} = \alpha_1 + \beta_1 \mathrm{Difi}_{i,t} + \beta_2 \mathrm{Fde}_{i,t} + \beta_3 \mathrm{Difi}_{i,t} * \mathrm{Fde}_{i,t} + \theta \mathrm{Control}_{i,t} + \mu_i + \tau_i + \varepsilon_{i,t} \tag{2}$$

Finally, to further investigate the nonlinear effects, using fiscal deficit as a threshold variable, the following model is constructed based on Hansen's panel threshold model:

$$\mathrm{LnPgdp}_{i,t} = \alpha_1 + \beta_1 \mathrm{Difi}_{i,t} * I(\mathrm{Fde}_{i,t} \le r) + \beta_2 \mathrm{Difi}_{i,t} * I(\mathrm{Fde}_{i,t} > r) + \theta \mathrm{Control}_{i,t} + \mu_i \tag{3}$$

In the above equation: μ_i and τ_t represents the fixed effects of region and time, $\varepsilon_{i,\,t}$ is a random perturbation term. In addition, I (·) is the indicative function, and r represents the threshold value to be estimated. When the conditions in parentheses are met, the indicative function can be expressed as 1, otherwise it is 0.

4 Empirical Analysis

4.1 Testing the Moderating Effect of Fiscal Deficits

The test results of the moderating effect of fiscal deficits are shown in Table 1.Column (2) indicates that the coefficient of Difi * Fde is 0.068, and through a 1% significance test, it indicates that fiscal deficits have a positive moderating effect on digital inclusive finance promoting economic growth. Hypothesis H2 is validated. Next, this article conducted the following robustness tests: firstly, incorporating the omitted variable of local fiscal revenue quality. Column (4) indicates that the coefficient of Difi * Fde is 0.055, which passes the 5% significance test and is similar to the empirical results mentioned above. The second is to change the sample size. Excluding the four municipalities directly under the central government of Beijing, Tianjin, Shanghai, and Chongqing, the sample will be regressed again. Column (6) indicates that the coefficient of Difi * Fde is 0.071, passing the 1% significance test. In summary, it can be considered that the empirical results are robust.

Table 1. Regulatory effects and robustness testing.

Variable name	Regulatory effects		Add missing variables		Change sample size	
	(1)	(2)	(3)	(4)	(5)	(6)
Difi	0.378***	0.445***	0.391***	0.443***	0.498***	0.571***
	(0.066)	(0.069)	(0.066)	(0.069)	(0.072)	(0.075)
Difi*Fde		0.068***		0.055**		0.071***
		(0.023)		(0.024)		(0.022)
Quality			−0.304***	−0.247***		
			(0.095)	(0.095)		
Fde	−1.515***	−1.283***	−1.495***	−1.311***	−1.406***	−1.149**
	(0.481)	(0.485)	(0.480)	(0.482)	(0.466)	(0.470)
cons	11.971***	12.012***	12.314***	12.283***	10.751***	10.862***
	(0.257)	(0.263)	(0.283)	(0.287)	(0.204)	(0.209)
N	372	372	372	372	324	324
Control vars	Y	Y	Y	Y	Y	Y
Double fixed	Y	Y	Y	Y	Y	Y
R^2	0.977	0.977	0.977	0.978	0.971	0.972
F	577.785	553.149	577.108	562.054	425.689	412.988

Note: * * *, * *, * * respectively represent significant levels at 1%, 5%, and 10%, with standard error in parentheses (the same below).

4.2 Heterogeneity Analysis

This article considers the following heterogeneity scenarios: One is different geographical regions. According to the classification criteria of previous literature, regressions were conducted for the eastern, central, and western regions, as shown in Table 2. It can be seen that in the eastern region, the coefficient of Difi * Fde is 0.693, passing the 1% significance test; In the western region, the coefficient of Difi * Fde is 0.078, passing the 5% significance test; In the central region, the coefficient of Difi * Fde is 0.248 and not significant. This indicates that the regulatory effect of fiscal deficits is superior in the eastern and western regions, while there is no effect in the central region, and hypothesis H2 can be verified. This may be due to the fact that the eastern region, relying on policy dividends and geographical advantages, has abundant financial resources such as human resources, trade, and capital. Its financial infrastructure construction is relatively complete, so the regulatory effect of fiscal deficits is strong. The western region has a vast area, low population density, and relatively remote geographical location, making it difficult for them to access high-quality financial resources, so the regulatory effect of fiscal deficits is weak. However, the central region is densely populated, with industries overly reliant on agriculture and mineral resource development. The technology and policy dividends of digital inclusive finance are lacking, making it difficult for fiscal deficits to exert regulatory effects.

Table 2. Heterogeneity test 1.

Variable name	Eastern region		Central region		Western region	
	(1)	(2)	(3)	(4)	(5)	(6)
Difi	−0.072	0.230^{***}	0.507^{***}	0.567^{***}	0.597^{***}	0.586^{***}
	(0.106)	(0.085)	(0.088)	(0.097)	(0.108)	(0.102)
Difi*Fde		0.693^{***}		0.248		0.078^{**}
		(0.099)		(0.216)		(0.033)
Fde	-4.530^{***}	-3.884^{***}	-2.457^{***}	-2.673^{***}	-0.530^{**}	−0.264
	(0.786)	(0.613)	(0.401)	(0.498)	(0.266)	(0.267)
cons	8.053^{***}	8.989^{***}	3.229^{**}	3.807^{**}	10.787^{***}	10.850^{***}
	(1.697)	(1.426)	(1.503)	(1.628)	(0.365)	(0.351)
N	132	132	96	96	144	144
Control vars	Y	Y	Y	Y	Y	Y
Double fixed	Y	Y	Y	Y	Y	Y
R^2	0.986	0.989	0.992	0.991	0.971	0.973

The second is the proportion of different people's livelihood fiscal expenditures. Regression was conducted based on the proportion of people's livelihood fiscal expenditure, as shown in Table 3. From it, it can be seen that when the proportion is low, the coefficient of Difi * Fde is 0.074. Through the 5% significance test, it indicates that the fiscal deficit has a positive regulatory effect; When the proportion is high, the coefficient of Difi * Fde is 0.070 and not significant, indicating that the fiscal deficit has no moderating effect, which is consistent with hypothesis H2. The reason for this phenomenon may be that when the proportion is low, fiscal funds usually invest in advanced technology fields, thus promoting the construction of financial infrastructure. At this time, the regulatory effect of fiscal deficits is strong. When the proportion is high, the fiscal deficit funds are mainly used to solve basic livelihood problems, and the investment in financial development is correspondingly reduced, so the fiscal deficit does not have a regulatory effect.

Table 3. Heterogeneity test 2.

Variable name	Low proportion of livelihood fiscal expenditure		High proportion of livelihood fiscal expenditure	
	(1)	(2)	(3)	(4)
Difi	0.247^{***}	0.502^{***}	0.366^{***}	0.424^{***}
	(0.086)	(0.109)	(0.102)	(0.121)
Difi*Fde		0.074^{**}		0.070
		(0.029)		(0.069)
Fde	-0.795^{***}	-1.961^{***}	-1.197^{**}	-1.030^{*}
	(0.265)	(0.474)	(0.524)	(0.557)
cons	12.564^{***}	12.198^{***}	11.851^{***}	11.850^{***}
	(0.405)	(0.402)	(0.429)	(0.429)
N	186	186	186	186
Control vars	Y	Y	Y	Y
Double fixed	Y	Y	Y	Y
R^2	0.984	0.977	0.977	0.977

4.3 Threshold Affect Test

The impact of digital inclusive finance on economic growth may have non-linear threshold conditions depending on the fiscal deficit being in different ranges. Therefore, this article uses fiscal deficit as a threshold variable for the following tests.

Based on Hansen's "Bootstrap" method, the F-value and P-value corresponding to the test statistic were obtained through repeated sampling of 300 times for threshold existence testing, as shown in Table 4. It can be seen that the P-values corresponding to the three threshold values and the double threshold values of fiscal deficit are 0.7133 and 0.1800, respectively, which are not significant. Next, we will conduct a single threshold estimation. At this point, the P-value is 0.0467, which passes the 5% significance test, indicating the existence of a single threshold effect. Then, further testing is shown in Table 5. The estimated threshold value of the single threshold model is 0.2137 and its corresponding 95% confidence interval is [0.2044, 0.2167].

Table 4. Threshold regression results.

Threshold variable	Number of thresholds	F-statistic	P-value	10%	5%	1%
Fde	Single	33.94**	0.0467	26.9289	32.9610	43.1926
	Double	17.13	0.1800	24.0757	29.1919	36.8525
	Three	8.02	0.7133	25.0004	29.9968	42.6220

Table 5. Threshold estimate and confidence interval.

Threshold variable	Threshold type	Threshold estimate	Confidence interval(95%)
Fde	Single threshold	0.2137	[0.2044, 0.2167]

Based on the above tests, threshold regression was performed as shown in Table 6. When the fiscal deficit is below the threshold value of 0.2137, the coefficient of digital inclusive finance on economic growth is 0.267; After crossing the threshold, the regression coefficient decreased to 0.220, and both passed the 1% significance test. This indicates that as the scale of fiscal deficits expands, the promoting effect of digital inclusive finance on economic growth exhibits a non-linear characteristic of first strong and then weak, verifying hypothesis H3. This may be due to the fact that when the fiscal deficit is within a reasonable range, it is conducive to the construction of financial infrastructure, financial industry, and digital technology innovation, so the promotion effect of digital inclusive finance is significant; But when the scale of the fiscal deficit is too large, it will cause a series of negative effects such as inflation, currency depreciation, and misallocation of scarce resources, and the promoting effect of digital inclusive finance will weaken.

Table 6. Threshold effect regression results.

Variable name	LnPgdp
Difi ($Fde \leq 0.2137$)	0.267^{***}
	(0.017)
Difi ($Fde > 0.2137$)	0.220^{***}
	(0.025)
cons	10.435^{***}
	(0.257)
N	372
Control vars	Y
R^2	0.892

5 Conclusion and Policy Recommendations

5.1 Conclusion

Firstly, fiscal deficits have a positive regulatory effect in promoting economic growth through digital inclusive finance.

Secondly, there is heterogeneity in the regulatory effect of fiscal deficits across different regions and the proportion of people's livelihood fiscal expenditures, manifested as being stronger in the eastern and western regions, while having no effect in the central region; When the proportion of livelihood fiscal expenditure is low, it is strong, but when the proportion is high, it has no effect.

Thirdly, fiscal deficit has a single threshold effect in promoting economic growth through digital inclusive finance, manifested as a non-linear characteristic where the promoting effect of digital inclusive finance on economic growth weakens when the fiscal deficit exceeds the threshold value.

5.2 Policy Suggestion

Firstly, tailored to local conditions, formulate differentiated fiscal deficits that are suitable for regional economic development. For the eastern region, it is recommended to implement a moderately proactive fiscal deficit, take advantage of the booming development of digital inclusive finance, and fully leverage the positive regulatory role of fiscal deficits. For the western region, it is recommended to implement a "proactive expansion" fiscal deficit. This move aims to increase fiscal investment, introduce financial resources and talents, strengthen regional coordination and cooperation, achieve cross regional integration of financial resources, and maximize the regulatory effect of fiscal deficits. For the central region, we should encourage the construction of financial infrastructure, increase construction efforts, and improve the level of construction; We need to actively promote the digital inclusive finance policy, enhance residents' financial literacy, enhance acceptance and participation, and thereby stimulate the potential regulatory effect of fiscal deficits.

Secondly, be meticulous in budgeting and focus on supporting digital inclusive finance projects to improve the efficiency of fiscal fund utilization. To fully utilize the effectiveness of fiscal funds, the following strategies need to be adopted. On the one hand, we must adhere to ensuring and improving people's livelihoods in development. On the basis of properly addressing people's livelihood issues, we will carry out precise management of fiscal funds, strictly control general expenditures, and ensure the rational distribution and efficient utilization of fiscal funds. By optimizing expenditure structure, we will strongly support the development of digital inclusive finance and related projects. On the other hand, consider appropriately expanding the scale of the fiscal deficit. Due to the basic fixed annual expenditure of local governments on people's livelihoods, appropriately expanding the scale of fiscal deficits will reduce their proportion, increase investment in advanced technology fields, accelerate the development of digital inclusive finance, and fully leverage the positive regulatory role of fiscal deficits.

Thirdly, controlling risks, setting reasonable upper limits for fiscal deficits, and achieving the optimal utility of digital inclusive finance. By reasonably setting the scale of fiscal deficits and promoting horizontal competition between regions, we can maximize the promoting effect of digital inclusive finance on economic growth. In addition, attention should also be paid to setting reasonable upper limits for fiscal deficits. On the basis of fully evaluating the economic situation, the government should prudently set the upper limit standards for fiscal deficits to ensure the rationality and effectiveness of fiscal policies. To prevent excessive fiscal deficits from disrupting economic stability and fiscal sustainability, and to curb local governments from shifting their fiscal funds towards areas with high short-term returns and quick results under enormous debt repayment pressure. By precisely regulating the scale of fiscal deficits, the government can not only mitigate risks but also fully leverage the promoting role of digital inclusive finance in economic growth.

Fund Project:. Philosophy and Social Science Fund of China "Research on the Dynamic Mechanism and Path of High-Quality Development of New Urbanization under the New Pattern of Double Circulation" (22BSH019).

References

1. Wang, Y., Pan, H.: Research on the sustainability conditions of fiscal deficits. Local Fin. Res. (05), 12–15+42 (2021)
2. Ouyang, H., Qin, S., Zhou, Z.: Can social institutional factors substantially affect fiscal deficits–Based on the test results of practical data from typical countries around the world. Publ. Finan. Res. (01), 50–63 (2022)
3. Qiao, C., Liu, Y., Yang, C.: The impact of FDI on agricultural technology innovation: an analysis of the regulatory effect of fiscal support for agriculture. Dongyue Luncong **44**(05), 148–159+192, (2023)
4. Ti, F.: An historic turning point in the adjustment of China's fiscal expenditure structure research on local finance. Livelihood Finan. (05), 4–8 (2008)
5. Wu, J.: Risk society and fiscal crisis: a sociological interpretation of the long term fiscal deficit. Southern Econ. (05), 17–29 (2023)
6. Guo, F., Wang, J., Wang, F., Kong, T., Zhang, X., Cheng, Z.: Measuring the development of digital inclusive finance in china: index compilation and spatial characteristics. Econ. Q., 1401–1418 (2020)
7. Wen, Z., Hou, J., Zhang, L.: Comparison and application of regulatory and mediating effects. J. Psychol. **37**(2), 268–74 (2005)

The Impact of E-Commerce on the Consumption Level of Chinese Residents Under the Background of Communication Network Technology

Xiaofeng Zhang and Weiyun Song(✉)

School of Public Finance and Administration, Harbin University of Commerce, Harbin 150028, Heilongjiang, China
2710125886@qq.com

Abstract. With the continuous development of modern communication technology and Internet technology, e-commerce standard in China is constantly improving. The thorough study of the influence of e-commerce on residents' consumption level is of great significance to enhance inhabitants' expenditure and stimulate economic progress. In the context of communication network technology, this article analyzes relevant literature, uses entropy method to measure the e-commerce level in China, and uses descriptive analysis and grey correlation analysis methods to conclude that e-commerce significantly affects the consumption level of Chinese residentst. Furthermore, it proposes countermeasures and suggestions to improve e-commerce development level and accelerate the improvement of consumer level among residents.

Keyword: Communication network technology · E-commerce · Resident consumption

1 Research Background

In the 1970s, China began to enter the era of informatization. With the fast development of global information skill, the level of communication network technology in China has also been continuously improving. The global mobile communication standard setting organization 3GPP pointed out at the 2024 Global 6G Technology Conference that the 6G service will complete the Rel-21 version standard in 2029. The development prospects of communication network skill are very broad, and the development speed is very fast. Driven by communication network technology, e-commerce scale in China has gradually expanded. In 2023, the market size of e-commerce in China reached 33.89 trillion, an increase of 7.92% compared to 2022. Meanwhile, e-commerce rapid progress is gradually affecting consumption standard of Chinese residents, gradually becoming an important way for their daily consumption.

X. Li et al. (Eds.): 6GN 2024, LNICST 615, pp. 209–218, 2025.
https://doi.org/10.1007/978-3-031-93193-2_18

2 Research Objective

The emerging e-commerce development model is gradually influencing residents' consumption behaviors, thereby affecting their consumption level. E-commerce may promote impulse consumption, and the live streaming e-commerce model in e-commerce can create an atmosphere of low prices and scarcity. Combined with the influence and language characteristics of influencers, it can stimulate consumers to place passionate orders and promote household consumption. This article mainly analyzes how e-commerce affects Chinese consumers' standard under the background of communication network technology through methods such as entropy method and grey correlation analysis, in order to provide suggestions for improving the consumption level of residents.

3 The Concept and Characteristics of E-Commerce

3.1 Current Research Status Abroad

Foreign research on the aspect primarily concentrate upon the impact of certain factors in e-commerce on household consumption and economic development.Harn A C P et al. (2006) used empirical analysis to analyze online shopping in Malaysia and found that online consumers in Malaysia lack confidence in e-commerce, which can inhibit the growth of consumption levels. Improvement is needed to enhance consumption levels [1].Intan Salwani et al. (2009) utilized the E-Value model to research how e-commerce affects tourism revenue, which can promote consumer consumption and increase tourism industry revenue [2].Yoo C W et al. (2013) pointed out that the good or bad electronic reputation in e-commerce can have an impact on residents' online shopping behavior, and positive reputation can motivate consumers to consume [3].Wilson et al. (2019) distributed 1000 questionnaires to consumers on five continents and used PLS-SEM method to research the significant impact of website design quality and service quality on consumer repurchase rates in e-commerce [4].Banda L G et al. (2023) collected time series data from the United States from 1990 to 2021 and used empirical analysis to conclude that e-commerce promotes household consumption, and there is a positive correlation between them [5].

3.2 Current Research Status in China

Domestic research on this aspect mainly focuses on how e-commerce affects the consumption level of rural residents in China,Li Junyang and Li Yaqi (2016) selected data related to e-commerce in their research, established a multiple regression model, and concluded that e-commerce could effectively stimulate inhabitants' expenditure standard growth [6].Liu Genrong (2017) analyzed the present situation of Chinese rural e-commerce progress, and based on this, studied how e-commerce affects Chinese consumer standard. Then, he analyzed the problems of rural e-commerce and proposed countermeasures and suggestions [7].Luo Jianping and Liu Jia (2018) conducted empirical analysis using panel data and concluded that e-commerce can promote the level of rural household consumption by raising their earning [8]. Wang Le (2022) used the system

GMM estimation method to analyze how e-commerce will promote the rural household consumption and enhance the rural economy [9]. Chen Xiao, Zhong Xiaoying, Huang Chun, et al. (2023) introduced the least squares method and multiple regression model for analysis, indicating that e-commerce will increase rural residents' income and improve their consumption level [10].

3.3 Review of Domestic and International Research

Through consulting relevant literature at home and abroad, it is concluded that foreign e-commerce started earlier, with a large number of literature and diverse research perspectives. However, domestic e-commerce started later, with a relatively small number of literature, and the research field still needs further expansion. This study further enriches the research system in related fields in China, which can be used as a reference for other researchers and provide reference for relevant departments to formulate policies.

4 Measurement of China's E-Commerce Level

Entropy method is an nonsubjective way of weighting, which determines weights through information entropy. Unlike AHP, entropy method is not easily influenced by subjective factors and can avoid some human interference. To study how e-commerce affects Chinese household consumpution level, the article collects relevant data through channels such as the CSMAR, and employs the entropy way to estimate the level of chinese e-commerce, thus obtaining results with higher objectivity and greater accuracy.

4.1 Review of Domestic and International Research

When using the entropy method to evaluate indicators, it is necessary to follow the principles of continuity, scientificity, and basis in the selection of indicator data, while constructing the basic framework of the indicator system.

Principles for Selecting Indicator Data. (1) Continuity. The selected indicator data is not intermittent, but continuous over a period of time. (2) Scientificity. The selected indicator data is correlated with the level of e-commerce development, which can comprehensively measure the indicator system. (3) Basis. This article draws on relevant literature and previous research results, selecting 10 indicators to measure the level of e-commerce in China, rather than subjective selection.

The Basic Framework of the Indicator System. Since relevant indicator information shortcoming before 2012, it may interfere with measurement. Therefore, so as to make certain the continuity, basis, and scientificity of indicator data, the article selects relevant data indicators of China's e-commerce development level from 2012 to 2022, and evaluates China's e-commerce level from ten indicators in four aspects: e-commerce related business, infrastructure, express delivery business, and mobile phone. Please refer to Table 1.

Table 1. Assessment indicator system for Chinese e-commerce.

Primary indicators	Secondary indicators	Third level indicators
E-commerce Level Index	E-commerce related business	Total Postal Business (X_1)
		Total telecommunications services (X_2)
	E-commerce infrastructure	Internet broadband access port (X_3)
		Internet penetration (X_4)
		Number of web pages (X_5)
	E-commerce express delivery business	Express delivery volume (X_6)
		Revenue from express delivery business (X_7)
	E-commerce mobile phone category	Mobile phone base station (X_8)
		Mobile phone users (X_9)
		Mobile phone penetration rate (X_{10})

4.2 Calculation of Indicators

Dimensionless Index. Supposing m samples and n evaluation indicators, the primeval index data matrix is $X = \left(x_{ij}\right)_{m\times n}(i = 1, 2 \ldots, m; j = 1, 2 \ldots, n)$. Among them, x_{ij} represents the value of the jth evaluation indicator in the i-th sample.

Due to the positive relationship between the ten indicators selected in this article and the level of e-commerce in China, when standardizing indicator data, only positive indicator processing method needs to be used, so only formula (1) needs to be inputted for processing.

$$x_{ij} = \frac{a_{ij} - mina_{ij}}{maxa_{ij} - mina_{ij}} \tag{1}$$

Then input the standardized data results into formula (2) for translation, with the translation amplitude T = 0.001.

$$y_{ij} = x_{ij} + T(0 < T \leq 1) \tag{2}$$

Entropy Value of Indicators. Insert the translated value into formula (3) and calculate the proportion of the j-th indicator in the i-th sample.

$$P_{ij} = \frac{x_{ij}}{\sum_{i=1}^{m} x_{ij}} \tag{3}$$

Substitute the calculation results of Pij into formula (4) and calculate the entropy value of each indicator e_j.

$$e_j = -\frac{1}{\ln m}[\sum_{i=1}^{m} p_{ij} \ln p_{ij}](j = 1, 2 \ldots, n) \tag{4}$$

Differential Coefficient of Indicators. Substitute the calculated entropy value e_j into formula (5) and calculate the coefficient of difference g_j for the indicator.

$$g_j = 1 - e_j \quad (5)$$

Weights of Indicators. Substitute the coefficient of difference g_j of the indicator into formula (6) and calculate the weight w_j of the indicator.

$$w_j = \frac{g_j}{\sum_{j=1}^{n} g_j} \quad (6)$$

Table 2. Entropy value, coefficient of difference, and weight of each evaluation indicator.

Number	Indicator Name	Entropy (e_j)	Coefficient of difference (g_j)	weight (w_j)
X1	Total Postal Business	0.8937	0.1063	0.1000
X2	Total telecommunications services	0.6734	0.3266	0.3071
X3	Internet broadband access port	0.9221	0.0779	0.0732
X4	Internet penetration	0.9218	0.0782	0.0735
X5	Number of web pages	0.9501	0.0499	0.0469
X6	Express delivery volume	0.8705	0.1295	0.1218
X7	Revenue from express delivery business	0.8915	0.1085	0.1020
X8	Mobile phone base station	0.9174	0.0826	0.0776
X9	Mobile phone users	0.9478	0.0522	0.0491
X10	Mobile phone penetration rate	0.9481	0.0519	0.0488

The Comprehensive Value Result Obtained By Linear Weighting Method. Substitute the translated indicator data results and the calculation results of indicator weight w_j into formula (7), and calculate the comprehensive value v_i of the linear weighting method (Table 3).

$$v_i = \sum_{j=1}^{n} (w_j \times y_{ij})(i = 1,2\ldots, m) \quad (7)$$

Table 3. Annual comprehensive value results.

Year	Comprehensive value (v_i)
2012	0.0001
2013	0.0595
2014	0.1192
2015	0.1882
2016	0.2476
2017	0.3491
2018	0.5332
2019	0.7263
2020	0.8877
2021	0.6289
2022	0.6688

4.3 The Results of the Equilibrium Index

Through the above methods to estimate the e-commerce in China, the weight results of various indicators in Table 2 are analyzed. It can be found that the total amount of telecommunications business, express business, express business income, and postal business have a relatively significant impact on the level of e-commerce in China. The impact of Internet broadband access ports, Internet penetration, mobile telephone base halt is relatively insignificant, the impact of the web pages quantity, mobel telephone customers, and mobile phone penetration is relatively insignificant. In other words, the role of e-commerce related business and express business is relatively large, while the role of e-commerce infrastructure and mobile phone is relatively small.

5 The Impact of E-Commerce on Chinese Inhabitants' Expenditure Level

To study how e-commerce influences Chinese inhabitants' expenditure level, the article chooses the method of grey correlation analysis for measurement. Grey correlation analysis is a multi factor statistical analysis method that has no limit on the sample size and can avoid discrepancies between quantitative and qualitative analysis results.

5.1 Analysis and Screening of Influencing Factors

This article mainly studies the correlation between e-commerce level (Z_1) and resident consumption level (Y). Therefore, the per capita consumption data of residents from the National Bureau of Statistics from 2012 to 2022 is taken as an indicator of the household consumption level, and the comprehensive value of e-commerce linearly weighted for each year is used as the level of e-commerce in China. In addition, four strong correlation factors that directly affect China's household consumption level, namely per capita income of residents (Z_2), per capita gross domestic product (Z_3), secondary and tertiary industries percentage (Z_4), employment in the tertiary industry percentage (Z_5), are selected as references. All four data are sourced from the 2012–2022 China Statistical Yearbook, and the data is continuous and authoritative.

5.2 Descriptive Analysis

In this section, Stata software was used to conduct descriptive analysis on the six indicators mentioned above. The logarithm of household consumption level, per capita income, and per capita gross domestic product were taken, and the comprehensive index of e-commerce level was obtained using the entropy method. A table of mean value, standard deviation, minimum and maximum values is obtained, see Table 4 for details. All six have relatively small standard deviations, the data volatility is not significant, and the degree of "variation" of the data is small. All indicators are relatively stable.

Table 4. Descriptive analysis.

Variable	Mean	Std. Dev	Min	Max
Resident consumption level (Y)	9.79348	0.23967	9.39713	10.10799
E-commerce level (Z_1)	0.40077	0.30201	0.0001	0.88771
Per capita income of residents (Z_2)	10.14745	0.26818	9.71169	10.51551
Per capita Gross Domestic Product (Z_3)	10.98398	0.25623	10.5909	11.35859
The proportion of the secondary and tertiary industries (Z_4)	0.92552	0.00911	0.91098	0.93703
The proportion of employment in the tertiary industry (Z_5)	0.43755	0.04032	0.361	0.48

5.3 Grey Correlation Analysis

Determine Reference Sequence and Comparison Sequence. Before conducting grey correlation analysis, the first step is to determine the reference sequence and comparison sequence. The reference sequence is the value of CPI from 2012 to 2022. Then, the data is subjected to dimensionless processing. In this paper, formula (8) is selected to standardize the initial values.

$$Y_i'(k) = \frac{Y_i(k)}{Y_i(1)} \tag{8}$$

Determine the Difference Sequence. The second step is to use formula (9) to determine the difference sequence.

$$\Delta_i(k) = \left|Y_1'(k) - Y_0'(k)\right| \tag{9}$$

Determine Maximum and Minimum Values. The third step is to use formulas (10) and (11) to determine the maximum and minimum values.

$$\Delta min = \min_{i}\left[\min_{k}\left|Y_l'(k) - Y_0'(k)\right|\right] \tag{10}$$

$$\Delta max = \max_{i}\left[\max_{k}\left|Y_l'(k) - Y_0'(k)\right|\right] \tag{11}$$

Calculate Correlation Coefficient. The fourth step is to use formula (12) to calculate the correlation coefficient, which is usually the resolution coefficient $\rho = 0.5$.

$$\xi_i(k) = \frac{\Delta min + \rho\Delta max}{\Delta_i(k) + \rho\Delta max} \tag{12}$$

Calculate Correlation Degree. The last step is to use formula (13) to calculate the correlation degree.

$$r_i = \frac{1}{n}\sum \xi_i(k) \tag{13}$$

5.4 Related Results

Table 5. Interdependence coefficients.

index	Z1	Z2	Z3	Z4	Z5
relational degree	0.595849	0.999984	0.999992	0.999885	0.99993

The results of grey relational analysis are shown in Table 5. The four variables Z_2, Z_3, Z_4, and Z_5 directly affect residents' consumption levels, so their correlation is very high, all reaching 0.9 or above. The per capita income and per capita gross domestic product of residents can to some extent reflect the level of national economic development and have a significant impact on residents' consumption. The proportion of the secondary and tertiary industries and the number of employees in the tertiary industry can to some extent reflect the level of development of the national industrial structure, and have a significant impact on residents' consumption. This article mainly analyzes how e-commerce level in China affects residents' consumption, with a correlation degree of nearly 0.6. Although the correlation degree is relatively low compared to the four variables Z_2, Z_3, Z_4, and Z_5, and the impact on residents' consumption is relatively small, generally, a correlation degree greater than 0.5 indicates a strong grey correlation. Therefore, based on the above results analysis, it can be said that e-commerce is an important factor affecting the consumption level of Chinese residents.

6 Conclusion and Suggestions for Countermeasures

6.1 Conclusion

Based on the above analysis, this article draws the following two conclusions. Firstly, the entropy method measure the e-commerce standard in our country by entropy method. It can be found that the total amount of telecommunications business has the most significant impact on e-commerce standard, the impact of express delivery business volume, express delivery business revenue, and total postal administration business is also relatively significant. It can be seen that telecommunications and express delivery transportation business are important factors affecting e-commerce standard and are closely related to e-commerce progress. Secondly, based on the results of grey correlation analysis, although the correlation between e-commerce and inhabitant expenditure is relatively low compared to indicators such as per capita income, per capita GDP, and employment rate in the tertiary industry, and the direct impact is relatively weak. However, the correlation index has reached a significant level, and it is necessary to be taken seriously.

6.2 Suggestions for Countermeasures

Gradually Enhance the E-commerce Level. On the one hand, as analyzed above, it can be seen that in the evaluation indicators of e-commerce level, the total amount of telecommunications services has the most significant impact. Therefore, the country can encourage the innovative development of satellite communication services and data communication services to improve the technological level of e-commerce. On the other hand, the government should first increase the construction of network and telecommunications infrastructure. Narrow the gap between the network and telecommunications infrastructure in the eastern, central and western regions, improve Internet penetration, improve Internet access services, and constantly expand the user base of e-commerce. Secondly, improve the circulation efficiency of the product. Related enterprises should promote express transportation business progress, accelerate the deep integration of modern message skill such as AI and datum analysis with express transportation business, enhance the productivity of express packaging and shipping, achieve real-time monitoring of express transportation, reduce labor costs, and gradually promote the modernization of express transportation business. Finally, cultivate high-quality talents in related fields. Strengthen talent cultivation internally, set up e-commerce training courses, and establish e-commerce majors in universities to fill the talent gap; Actively introducing high-quality talents in this field to the outside world, providing a continuous supply of talent for improving the e-commerce level.

Raising Residents' Consumption Level. On the one hand, promoting national economic development is of paramount importance in improving residents' consumption levels. The improvement of the national economic development level can drive the growth of per capita GDP and per capita income of residents, as well as promote the development of the secondary and tertiary industries, thereby increasing the demand for labor in related industries and having a direct impact on promoting the improvement of residents' consumption level. On the other hand, relevant enterprises should actively expand multi-channel e-commerce development methods. First of all, enterprises should make good use of the publicity and promotion role of social media such as Tiktok and Xiaohongshu, and transform the "people looking for goods" model into the "goods looking for people" model through live broadcast e-commerce, video promotion and other ways to actively develop social e-commerce; Secondly, enterprises should actively explore innovative new retail e-commerce operation models that integrate online and offline development, perfect various aspects of product production, circulation, and sales, and achieve omnichannel marketing; Finally, enterprises should develop cross-border e-commerce to meet consumers' various shopping needs and provide them with a brand new shopping experience. Continuously promoting the modernization and digitization of e-commerce, perfecting related infrastructure, and continuously enhancing residents' consumption level.

Fund Project:. Philosophy and Social Science Fund of China "Research on the Dynamic Mechanism and Path of High-Quality Development of New Urbanization under the New Pattern of Double Circulation" (22BSH019).

References

1. Harn, A.C.P., Khatibi, A., Ismail, H.B.: E-commerce: a study on online shop** in Malaysia. J. Soc. Sci. **13**(3), 231–242 (2006)
2. Intan Salwani, M., et al.: E-commerce usage and business performance in the Malaysian tourism sector: empirical analysis. Inf. Manag. Comput. Secur. **17**(2), 166–185 (2009)
3. Yoo, C.W., Sanders, G.L., Moon, J.: Exploring the effect of e-WOM participation on e-Loyalty in e-commerce. Decis. Support Syst. **55**(3), 669–678 (2013)
4. Wilson, N., Keni, K., Tan, P.H.P.: The effect of website design quality and service quality on repurchase intention in the e-commerce industry: a cross-continental analysis. Gadjah Mada Int. J. Bus. **21**(2), 187–222 (2019)
5. Banda, L.G., Kassam, Z.Z.: E-commerce and household consumption in the United States: an arrangement of convenience. Cogent Bus. Manage. **10**(3), 2275360 (2023)
6. Li, J., Li, Y.: Empirical study on the impact of E-commerce on consumption. J. Tianjin Univ. Commer. **36**(04), 23–28+63 (2016)
7. Genrong, L.: Analysis of the mechanism of the impact of E-commerce on rural residents' consumption. China Circ. Econ. **31**(05), 96–104 (2017)
8. Luo, J., Liu, J.: Research on the impact of E-commerce on rural residents' consumption: an intermediary effect analysis based on inter provincial panel data. Price Theory Pract. (05), 131–134 (2018)
9. Le, W.: Research on the impact of E-commerce on the consumption level of rural residents. Agric. Technol. **42**(16), 172–175 (2022)
10. Xiao, C., et al.: Empirical study on the impact of E-commerce on rural residents' consumption. Agric. Econ. **03**, 134–135 (2023)

Author Index

C
Chen, Dawei 113
Chen, Jiantian 179
Cheng, Luxi 153

D
Deng, Xiang 83

F
Fan, Qilin 83, 98

G
Guo, Jiajie 113

H
Hao, Jinlong 71, 153
He, Jian 113
Hou, Zufeng 179
Huang, Tao 14
Huo, Ru 14

J
Ji, Fangfang 199

L
Li, Ao 169, 191
Li, Chenyu 3
Li, Han 55
Li, Haowei 98
Li, Hui 153
Li, Jingzhong 134
Li, Penghua 71, 153
Li, XinLu 125
Li, Xiuhua 71, 83, 98, 153
Lin, XiuLian 125
Liu, Haixin 179
Liu, Tong 45
Liu, Zhiwei 169, 191
Lu, Jiangang 179
Lu, Wanyu 169, 191

M
Ming, Zihong 83

P
Pan, Kaiyan 179

S
Sha, Zongxuan 14
Shi, Ruohan 83
Shi, Shuo 3, 45, 55, 113
Song, Tao 153
Song, Weiyun 209
Sui, Shangwen 169, 191
Sun, Jialei 3

T
Tang, Jian 71

W
Wang, Chao 179
Wang, Gang 71
Wang, Kai 83
Wang, Xi 3
Wei, Shouming 55
Wu, Peiran 28

X
Xiong, Zhiying 98
Xu, Guozeng 71
Xu, Yinkang 98
Xu, Zhenyu 125, 134

Y
Ye, Haiheng 28
Yin, Zhongyu 45

X. Li et al. (Eds.): 6GN 2024, LNICST 615, pp. 219–220, 2025.
https://doi.org/10.1007/978-3-031-93193-2

Yu, Bin 169, 191

Z
Zhang, Cheng 55, 98
Zhang, Kexin 45
Zhang, Xiaofeng 199, 209
Zhang, Yanhong 134
Zhang, Yong 179
Zhao, Ruifeng 179
Zhenduo, Fan 143

The manufacturer's authorised representative in the EU is Springer Nature Customer Service Centre GmbH, Europaplatz 3, 69115 Heidelberg, Germany. If you have any concerns regarding our products, please contact ProductSafety@springernature.com

Printed and bound by CPI Group (UK) Ltd, Croydon, CR0 4YY
01/10/2026
02225908-0003